Intelligent Workloads at the Edge

Deliver cyber-physical outcomes with data and machine learning using AWS IoT Greengrass

Indraneel Mitra

Ryan Burke

BIRMINGHAM—MUMBAI

Intelligent Workloads at the Edge

Copyright © 2022 Packt Publishing

Publishing Product Manager: Reshma Raman
Senior Editor: Roshan Kumar
Content Development Editor: Tazeen Shaikh
Technical Editor: Rahul Limbachiya
Copy Editor: Safis Editing
Project Coordinator: Aparna Ravikumar Nair
Proofreader: Safis Editing
Indexer: Sejal Dsilva
Production Designer: Shyam Sundar Korumilli
Marketing Coordinator: Nimisha Dua

First published: January 2022

Production reference: 1101221

Published by Packt Publishing Ltd.
Livery Place
35 Livery Street
Birmingham
B3 2PB, UK.

ISBN 978-1-80181-178-1

www.packt.com

To Guruji and the Almighty for bestowing immense blessings and grace on my life. To my mom and dad for all their love, support, and sacrifices and for making me the person I am today. To my wife, for her continued support of my dreams and for being my loving partner throughout our life journey. To my children, Ivan and Aayan, for bringing enormous joy, determination, and enthusiasm to my life. And to my in-laws for the immense support they have given to my family.

To my teachers, especially Ashok Mitra, Utpal Basu, and Gautam Mahapatra, for giving direction to my life. To my peers turned lifelong friends who have believed in me and stood by my crazy ideas over the test of time – Sudip Dasgupta, Rudra Bhattacharjee, Amit Mukherjee, and Harish Rajagopalan.

I also want to share my heartfelt thanks to the AWS IoT and Automotive Technical Field community, along with the incredible leaders who have mentored me over the years: Olawale Oladehin, Gavin Adams, Dean Phillips, Nisarg Modi, Kanwal Duggal, and Brett Francis.

I also want to express my sincere gratitude to James Gosling and coauthor Ryan Burke for making this book a reality. And finally, to the technical reviewers and collaborators – Gavin Adams, Ashok Rao, and Amit Mukherjee – for their ideas and feedback to help refine and polish the content of this book.

– Indraneel (Neel) Mitra

To my wife, Nicole, for supporting me while taking on a book project in addition to working full time. To my friends and family for their support during this time. And special thanks to the AWS IoT Technical Field community and the leaders in my career at Amazon who have helped me to grow: Rahul Sareen, Olawale Oladehin, Glenn Dasmalchi, Sarah Cooper, Vadim Jelezniakov, Shyam Krishnamoorthy, and Kyle Roche.

– Ryan Burke

Contributors

About the authors

Indraneel Mitra is a Principal Solutions Architect for IoT at Amazon Web Services. He has 17+ years of IT consulting and architecting experience with start-ups and Fortune 500 customers across different industry verticals. He has spoken at different events, including AWS re:Invent, AWS Summits, AWS Webinar and Twitch shows, the RSA cybersecurity conference, the **Emerging Technologies for Enterprise (ETE)** conference, and local meetups on different topics for many years as a seasoned public speaker. Neel is also the co-author and trainer of the popular IoT course, *AWS IoT: Developing and Deploying an Internet of Things*, available on Coursera and edX. He holds an M.S. in software engineering from BITS Pilani, India.

Ryan Burke is a Senior Sustainability Application Architect at Amazon Web Services, formerly the Worldwide Technical Leader for IoT. He has worked in information technology since 2006, developing web services, user experiences, and architecting IoT solutions. He has spoken to audiences on IoT topics at events including AWS re:Invent, SxSW, TDWI, and local meetup communities. He holds a B.S. in computer science from the Georgia Institute of Technology, and served 6 years as a communications officer in the U.S. Air Force Reserves. In his free time, Ryan enjoys yoga, coffee snobbery, games of all kinds, and annoying his family with a new smart home project.

About the reviewers

Ashok Rao is an IoT specialist solutions architect at AWS. He has a master's degree in electronics engineering and has been working with many global customers in the embedded systems and IoT domains for the past 10+ years. Ashok has helped design, develop, and deploy many solutions and advise companies as they navigate the challenges of moving IoT projects from concept to production. He also has experience working with microcontroller devices, RTOSes, and firmware development for various devices in the consumer electronics domain. This gives him a rare mix of knowledge both in the device/hardware layer and cloud computing spaces.

I would like to thank my parents and son, Abhay Rao, for the opportunity to contribute to this book.

Gavin Adams is a Principal Solutions Architect with Amazon Web Services, specializing in helping customers deploy IoT solutions. Prior to his current role at AWS, Gavin began his technology career supporting nuclear energy testing missions in the Nevada desert. Since then, he has been involved in other aspects of information technology, including software development, security, system automation, system architecture, and design, and since 2012, he has focused primarily on cloud-native solutions for customers. He now focuses on the delivery of edge compute solutions, putting him back in the physical world with hands-on systems leveraging cloud capabilities, and sometimes, even back in those same arid landscapes.

Table of Contents

3
Building the Edge

4
Extending the Cloud to the Edge

5

Ingesting and Streaming Data from the Edge

6

Processing and Consuming Data on the Cloud

7
Machine Learning Workloads at the Edge

Section 3: Scaling It Up

8
DevOps and MLOps for the Edge

9

Fleet Management at Scale

Section 4: Bring It All Together

10

Reviewing the Solution with AWS Well-Architected
Framework

Preface

The **Internet of Things (IoT)** has transformed how businesses think about and interact with the world. Sensors can measure the performance of high-volume industrial manufacturing operations or the daily environmental health of a remote island. The IoT makes it possible to study the world at various levels of precision and enable data-driven decision making anywhere. **Machine learning (ML)** and Elastic cloud computing have accelerated our ability to understand and analyze the huge amount of data generated by the IoT. With edge computing, data analytics and ML models can process information locally at the source where the data is generated.

This book will teach you to combine the technologies of edge computing and machine learning to deliver next-generation cyber-physical outcomes. You'll begin by discovering how to create software applications that run on edge devices using software from Amazon Web Services, such as AWS IoT Greengrass. As you advance, you'll learn how to process and stream IoT data from the edge to the cloud and use it to train ML models using Amazon SageMaker. The book also shows you how to optimize these models and run them at the edge for optimal performance, cost savings, and data compliance.

By the end of this book, you'll be able to scope your own IoT workloads, bring the power of machine learning to the edge, and operate those workloads in a production setting.

Who this book is for

This book is for IoT architects and software engineers responsible for delivering analytical and machine learning-backed software solutions to the edge. Amazon Web Services users who want to learn and build IoT solutions will find this book useful. Intermediate experience with running Python software on Linux is required to make the most of this book.

What this book covers

Chapter 1, Introduction to the Data-Driven Edge with Machine Learning, introduces concepts such as the edge and how machine learning has a unique value when run at the edge. It provides a high-level overview of use cases in consumer and industrial settings. It sets the context for the fictional scenario that will guide hands-on activities for the book.

Chapter 2, Foundations of Edge Workloads, provides an overview of key considerations for designing edge solutions and includes an introduction to the use of AWS IoT Greengrass.

Chapter 3, Building the Edge, dives into next-level details of building edge solutions through the more advanced use of AWS IoT Greengrass to author software components for your business logic.

Chapter 4, Extending the Cloud to the Edge, introduces how to build edge solutions with native cloud connectivity and deploy software to remote devices over the internet. It also introduces software components provided by AWS for abstracting away common needs for edge functionality.

Chapter 5, Ingesting and Streaming Data from the Edge, introduces how to perform data modeling for IoT workloads and why it's important. It also introduces various architectural patterns and anti-patterns for collecting, ingesting, and processing data streams on the edge.

Chapter 6, Processing and Consuming Data on the Cloud, explains how the integration of IoT with big data technologies enables high-volume complex data processing in the cloud. It also dives deeper into how to extend the data processing design patterns from the edge to the cloud to unblock advanced use cases.

Chapter 7, Machine Learning Workloads at the Edge, introduces the concepts of machine learning in the context of IoT workloads. It also dives deeper into the different phases of machine learning workflow along with applicable design patterns and anti-patterns.

Chapter 8, DevOps and MLOps for the Edge, explains how the concepts of DevOps and MLOps can be leveraged for IoT workloads to enable agile development practices from the cloud to the edge.

Chapter 9, Fleet Management at Scale, introduces the concepts of fleet management using cloud-native IoT toolchains. It also dives deeper into the different scenarios and mechanisms applicable for onboarding IoT devices at scale in the real world.

Chapter 10, Reviewing the Solution with AWS Well-Architected Framework, concludes the book with a synopsis of key lessons and steps in terms of how to approach reviewing a solution's design with a multi-faceted review framework from AWS. It also offers ideas on the next steps for IoT architects to take given the lessons learned from this book.

To get the most out of this book

You will need a personal computer running Windows, macOS, or Linux. This computer uses the AWS Command Line Interface in a terminal and the AWS Management Console through a web browser. A second, Linux-based system acts as the edge device and hosts the edge solution running AWS IoT Greengrass. This second system can be a local or remote virtual machine or an actual device like a Raspberry Pi. For the real IoT experience, we recommend using a Raspberry Pi 3B (or later) with a SenseHAT expansion board to complete the hands-on portions of the book. If you do not have a hardware device, you can use an Ubuntu Linux virtual machine instead. Ultimately, you can finish all hands-on steps with or without a second device.

The use of AWS for cloud-based services does incur a small cost. You will need access to an AWS account or create one yourself. Completion of all hands-on sections can accrue billing up to $25 US Dollars (USD). You can opt out of the ML training steps to reduce the cost to less than $5.

Software/hardware covered in the book	Operating system requirements
AWS account and AWS Command Line Interface	Windows, macOS, or Linux
Python 3.7+	Windows, macOS, or Linux
AWS IoT Greengrass v2	Linux
AWS IoT Core, Amazon DynamoDB, Amazon Simple Storage Service, AWS Glue, Amazon SageMaker, Amazon Athena, Amazon QuickSight	N/A (cloud-based services)

At the time of authoring, AWS IoT Greengrass v2 did not support Windows installation. The hands-on portions related to the edge solution are specific to Linux and do not run on Windows.

If you are using the digital version of this book, we advise you to type the code yourself or access the code from the book's GitHub repository (a link is available in the next section). Doing so will help you avoid any potential errors related to the copying and pasting of code.

Download the example code files

You can download the example code files for this book from GitHub at `https://github.com/PacktPublishing/Intelligent-Workloads-at-the-Edge`. If there's an update to the code, it will be updated in the GitHub repository.

We also have other code bundles from our rich catalog of books and videos available at `https://github.com/PacktPublishing/`. Check them out!

Download the color images

We also provide a PDF file that has color images of the screenshots and diagrams used in this book. You can download it here:

`https://static.packt-cdn.com/downloads/9781801811781_ColorImages.pdf`

Conventions used

There are a number of text conventions used throughout this book.

`Code in text`: Indicates code words in the text, database table names, folder names, filenames, file extensions, pathnames, dummy URLs, user input, and Twitter handles. Here is an example: "To use Amazon SageMaker Debugger, you must enhance `Estimator` with three additional configuration parameters: `DebuggerHookConfig`, `Rules`, and `ProfilerConfig`."

A block of code is set as follows:

```
#Feature group name
weather_feature_group_name_offline = 'weather-feature-group-
offline' + strftime('%d-%H-%M-%S', gmtime())
```

When we wish to draw your attention to a particular part of a code block, the relevant lines or items are set in bold:

```
@smp.step
def train_step(model, data, target):

        output = model(data)
        long_target = target.long()
        loss = F.nll_loss(output, long_target, reduction="mean")
        model.backward(loss)
        return output, loss
    return output, loss
```

Any command-line input or output is written as follows:

```
$ git clone PacktPublishing/Intelligent-Workloads-at-the-Edge-
```

Bold: Indicates a new term, an important word, or words that you see on screen. For instance, words in menus or dialog boxes appear in **bold**. Here is an example: Keep in mind that when you use multiple instances in the training cluster, all instances should be in the same **Availability Zone**.

> **Tips or Important Notes**
> Appear like this.

Get in touch

Feedback from our readers is always welcome.

General feedback: If you have questions about any aspect of this book, email us at customercare@packtpub.com and mention the book title in the subject of your message.

Errata: Although we have taken every care to ensure the accuracy of our content, mistakes do happen. If you have found a mistake in this book, we would be grateful if you would report this to us. Please visit www.packtpub.com/support/errata and fill in the form.

Piracy: If you come across any illegal copies of our works in any form on the internet, we would be grateful if you would provide us with the location address or website name. Please contact us at copyright@packt.com with a link to the material.

If you are interested in becoming an author: If there is a topic that you have expertise in and you are interested in either writing or contributing to a book, please visit authors.packtpub.com.

Share Your Thoughts

Once you've read *Intelligent Workloads at the Edge*, we'd love to hear your thoughts! Scan the QR code below to go straight to the Amazon review page for this book and share your feedback.

https://packt.link/r/1-801-81178-4

Your review is important to us and the tech community and will help us make sure we're delivering excellent quality content.

Section 1: Introduction and Prerequisites

This section will introduce you to common edge concepts, personas, challenges, and tools used to deliver edge outcomes. This section will also introduce you to the project that you'll build over the course of the hands-on chapters.

This section comprises the following chapter:

- *Chapter 1, Introduction to the Data-Driven Edge with Machine Learning*

1

Introduction to the Data-Driven Edge with Machine Learning

The purpose of this book is to share prescriptive patterns for the **end-to-end** (E2E) development of solutions that run at the **edge**, the space in the computing topology nearest to where the analog interfaces the digital and vice versa. Specifically, the book focuses on those edge use cases where **machine learning** (ML) technologies bring the most value and teaches you how to develop these solutions with contemporary tools provided by **Amazon Web Services** (AWS).

In this chapter, you will learn about the foundations for cyber-physical outcomes and the challenges, personas, and tools common to delivering these outcomes. This chapter briefly introduces the smart home and industrial **internet of things** (**IoT**) settings and sets the scene that will steer the hands-on project built throughout the book. It will describe how ML is transforming our ability to accelerate decision-making beyond the cloud. You will learn about the scope of the E2E project that you will build using AWS services such as **AWS IoT Greengrass** and **Amazon SageMaker**. You will also learn what kinds of technical requirements are needed before moving on to the first hands-on chapter, *Chapter 2, Foundations of Edge Workloads*.

The following topics will be covered in this chapter:

- Living on the edge
- Bringing ML to the edge
- Tools to get the job done
- Demand for smart home and industrial IoT
- Setting the scene: A modern smart home solution
- Hands-on prerequisites

Living on the edge

The **edge** is the space of computing topology nearest to where the analog interfaces the digital and vice versa. The edge of the first computing systems, such as 1945's **Electronic Numerical Integrator and Computer** (**ENIAC**) general-purpose computer, was simply the interfaces used to input instructions and receive printed output. You couldn't access these interfaces without being directly in front of them. With the advent of remote access mainframe computing in the 1970s, the edge of computing moved further out to public terminals that fit on a desk and connected to mainframes via coaxial cable. Users could access the common resources of the local mainframe from the convenience of a lab or workstation to complete their work with advanced capabilities such as word processors or spreadsheets.

The evolution of humans using the edge for computing continued with increases in compute power and decreases in size and cost. The devices we use every day, such as personal computers and smartphones, deliver myriad outcomes for us. Some outcomes are delivered entirely at the edge (on the device), but many work only when connected to the internet and consume remote services. Edge workloads for humans tend to be diverse, multipurpose, and handle a range of dynamisms. We could not possibly enumerate everything we could do with a smartphone and its web browser! These examples of the edge all have in common that humans are both the operator and recipient of a computing task. However, the edge is more than the interface between humans and silicon.

Another important historical trend of the edge is autonomous functionality. We design computing machines to sense and act, then deploy them in environments where there may be no human interaction at all. Examples of the autonomous edge include robotics used in manufacturing assembly, satellites, and weather stations. These edge workloads are distinct from human-driven workloads in that they tend to be highly specialized, single-purpose, and handle little dynamism. They perform a specific set of functions, perform them consistently, and repeat them until obsolescence. The following figure provides a simplistic history of both human-driven interfaces and autonomous machines at the edge over time:

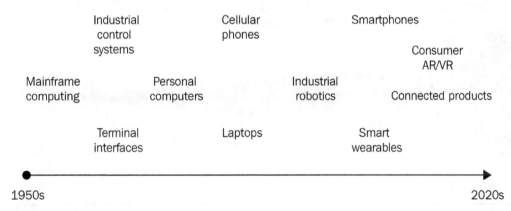

Figure 1.1 – A timeline of cyber-physical interfaces at the edge from 1950 to 2020

In today's technological advances of wireless communications, microcontrollers and microprocessors, electrical efficiency, and durability, the edge can be anywhere and everywhere. Some of you will be reading this book on an e-reader, a kind of edge device, at 10 **km** altitude, cruising at 900 **kph**. The Voyager 1 spacecraft is the most distant manmade edge solution, continuing to operate at the time of this writing 152 **AU** from Earth! The trend here is that over time, the spectrum of capabilities along the path to and at the edge will continue to grow, as will the length of that path (and the number of points on it!) and the remoteness of where those capabilities can be deployed. The following diagram illustrates the scaling of entities, compute power, and capabilities across the topology of computing:

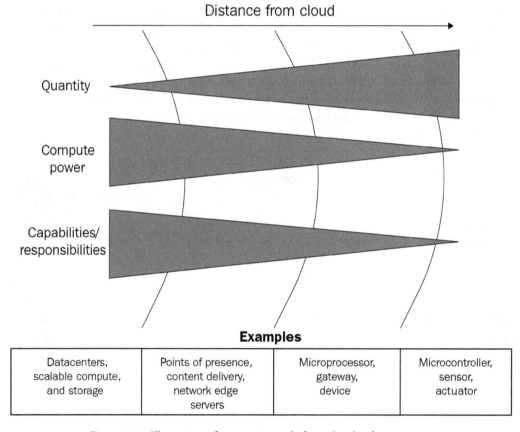

Figure 1.2 – Illustration of computing scale from the cloud to a sensor

Our world is full of sensors and actuators; over time, more of these devices are joining the IoT. A **sensor** is any component that takes a measurement from our analog world and converts it to digital data. An **actuator** is any component that accepts some digital command and applies some force or change out into the analog world. There's so much information out there to collect, reason about, and act upon. Developing edge solutions is an exciting frontier for the following reasons:

- There is a vast set of possibilities and problems to solve in our world today. We need more innovation and solutions to address global outcomes, such as the 17 sustainable development goals defined by the **United Nations** (**UN**).

- The shrinking cost factor to develop edge solutions lowers the barrier to experiment.

- Tools that put solution development in the reach of anyone with a desire to learn are maturing and becoming simpler to use.

This book will teach you how to develop the software of edge solutions using modern edge-to-cloud technologies, including how to write software that interacts with physical sensors and actuators, how to process and exchange data with other local devices and the cloud, and how to get value from advanced ML technologies at the edge. More important than the how is the why—in other words: *why do we build the solutions this way?* This book will also explain the architectural patterns and tips for building well-architected solutions that will last beyond the time of particular technologies and tools.

Implementation details such as programming languages and frameworks come and go with popularity, necessity, and technological breakthroughs. The patterns of what we build and why we build them in particular ways stand the test of time and will serve you for many of your future projects. For example, the 1995 *Design Patterns: Elements of Reusable Object-Oriented Software* by *Gamma, Helm, Johnson,* and *Vlissides* is still guiding software developers today despite the evolution of tools that the authors used at the time. We, the authors, cannot liken ourselves to these great thinkers or their excellent book, but we refer to it as an example of how we approached writing this book.

Common concepts for edge solutions

For the purposes of this book, we will expand the definition of the edge as any component of a cyber-physical solution operating outside of the cloud, its data centers, and away from the internet backbone. Examples of the edge include a radio switch controlling a smart light bulb in a household, sensors recording duty cycles and engine telemetry of a tractor-trailer at a mining site, a turnstile granting access to subway commuters, a weather buoy drifting in the Atlantic, a smartphone using a camera in a new **augmented reality (AR)** game, and of course, Voyager 1. The environmental control system running in a data center to keep servers cool is still an edge solution; our definition intends to highlight those components that are distant from the *gravity* of the worldwide computing topology. The following diagram shows examples of computing happening at various distances further from the gravity of data centers in this computing topology:

Figure 1.3 – Examples of the edge at various distances

A **cyber-physical solution** is one that *combines hardware and software for interoperating the digital world with the analog world*. If the analog world is a set of properties we can measure about reality and enact changes back upon it, the digital world is the information we capture about reality that we can store, transmit, and reason about. A cyber-physical solution can be self-contained or a closed loop such as a **programmable logic controller (PLC)** in an industrial manufacturing shop or a digital thermostat in the home. It may perform some task autonomously or at the direction of a local actor such as a person or switch.

An **edge solution** is, then, *a specific extension of a cyber-physical solution in that it implies communication or exchange of information with some other entity at a point in time*. It can also operate autonomously, at the direction of a local actor or a remote actor such as a web server. Sensors deliver autonomous functionality in the interest of a person who needs the provided data or are used as input to drive a decision through a PLC or code running on a server. Those decisions, whether derived by computers or people, are then enacted upon through the use of actuators. The analog, human story of reacting to the cold is *I feel cold, so I will start a fire to get warm*, while the digital story might look like the following pseudocode running a local controller to switch on a furnace:

```
if ( io.read_value(THERMOMETER_PIN) < LOWER_HEAT_THRESHOLD ) {
    io.write_value(FURNACE_PIN, HIGH);
}
```

We will further define edge solutions to have some compute capability such as a microcontroller or microprocessor to execute instructions. These have at least one sensor or actuator to interface with the physical world. Edge solutions will at some point in time interact with another entity on a network or in the physical world, such as a person or another machine.

Based on this definition, what is the edge solution that is nearest to you right now?

Sensors, actuators, and compute capability are the most basic building blocks for an edge solution. The kinds of solutions that we are interested in developing have far more complexity to them. Your responsibility as an IoT architect is to ensure that edge solutions are secure, reliable, performant, and cost-effective. The high bar of a well-architected solution means you'll need to build proficiency in fundamentals such as networking, cryptography, electrical engineering, and operating systems. Today's practical production edge solutions incorporate capabilities such as processing real-time signals, communicating between systems over multiple transmission media, writing firmware updates with redundant failure recovery, and self-diagnosing device health.

This can all feel overwhelming, and the reality is that building solutions for the edge is both complex and complicated. In practice, we use purpose-built tools and durable patterns to focus invested efforts on innovation and problem solving instead of bootstrapping and reinventing the wheel. The goal of this book is to start small with functional outcomes while building up to the big picture of an E2E solution. The learnings along the way will serve you on your journey of building your next solution. While this book doesn't cover every topic in the field, we will take every opportunity to highlight further educational resources to help you build proficiency beyond the included focus areas. And that's just about all in terms of building solutions for the edge! Next, let's review how ML fits in.

Bringing ML to the edge

ML is an incredible technology making headway in solving today's problems. The ability to train computers to process great quantities of information in service of classifying new inputs and predicting results rivals, and in some applications exceeds, what the human brain can accomplish. For this reason, *ML defines mechanisms for developing artificial intelligence (AI)*.

The vast computing power made available by the cloud has significantly reduced the amount of time it takes to train ML models. Data scientists and data engineers can train production models in hours instead of days. Advances in ML algorithms have made the models themselves ever more portable, meaning that running the models can work on computers with smaller compute and memory profiles. The implications of delivering portable ML models cannot be overstated.

Operating ML models at the edge helps us as architects deliver optimal edge solution design principles. By hosting a portable model at the edge, the proximity to the rest of our solution leads to four key benefits, outlined as follows:

- First, this means the solution can maximize responsiveness for capabilities depending on the results of ML inferences by not waiting for the round-trip latency of a call to a remote server. The latency to interpret myriad signals from an engine about to fail can be made in 10 **milliseconds (ms)** instead of 100 ms. This degree of latency can make the difference between a safe operation and a catastrophic failure.

- Second, it means the functionality of the solution will not be interrupted by network congestion and can run in a state where the edge solution is disconnected from the public internet. This opens up possibilities for ML solutions to run untethered from cloud services. That imminent engine failure can be detected and prevented regardless of connection availability.

- Third, anytime we can process data locally with an ML model and reduce the quantity of data that ultimately needs to be stored in the cloud, we also get the cost-saving benefits on transmission. Think of an expensive satellite internet provider contract; across that kind of transmission medium, IoT architects only want to transmit data that is absolutely necessary to keep costs down.

- Fourth, another benefit of local data processing is that it enables use cases that must conform to regulation where data must reside in the local country or observe privacy concerns such as healthcare data. Hospital equipment used to save lives arguably needs as much intelligent monitoring as it can get, but the runtime data may not legally be permitted to leave the premises.

These four key benefits are illustrated in the following diagram:

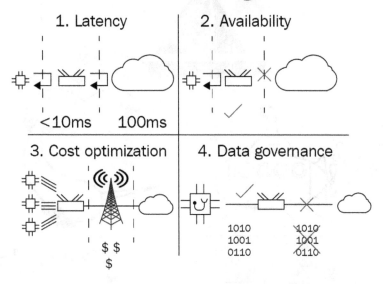

Figure 1.4 – The four key benefits of ML at the edge

Imagine a submersible drone that can bring with it an ML model that can classify images coming from a video feed. The drone can operate and make inferences on images away from any network connection and can discard any images that don't have any value. For example, if the drone's mission is to bring back only images of narwhals, then the drone doesn't need extensive quantities of storage to save every video clip for later analysis. The drone can use ML to classify images of narwhals and only preserve those for the trip back home. The cost of storage continues to drop over time, but in the precious bill of materials and space considerations of edge solutions such as this one, bringing a portable ML model can ultimately lead to significant cost savings.

The following diagram illustrates this concept:

Figure 1.5 – Illustration of a submersible drone concept processing photographs and storing only those where a local ML model identifies a narwhal in the subject

This book will teach you the basics of training an ML model from the kinds of machine data common to edge solutions, as well as how to deploy such models to the edge to take advantage of combining ML capabilities with the value proposition of running at the edge. We will also teach you about operating ML models at the edge, which means analyzing the performance of models, and how to set up infrastructure for deploying updates to models retrained in the cloud.

Outside the scope of this book's lessons are comprehensive deep dives on the data science driving the field of ML and AI. You do not need proficiency in that field to understand the patterns of ML-powered edge solutions. An understanding of how to work with **input/output (I/O)** buffers to read and write data in software is sufficient to work through the ML tools used in this book.

Next, let's review the kinds of tools we need to build and the specific tools we will use to build our solution.

Tools to get the job done

This book focuses on tools offered by AWS to deliver ML-based solutions at the edge. Leading with the 2015 launch of the AWS IoT Core service, AWS has built out a suite of IoT services to help developers build cyber-physical solutions that benefit from the power of the cloud. These services range from edge software, such as the FreeRTOS real-time operating system for microcontrollers, to **command and control (C2)** of device fleets with IoT Core and IoT Device Management, and analytical capabilities for yielding actionable insights from data with services such as IoT SiteWise and IoT Events. The IoT services interplay nicely with Amazon's suite of ML services, enabling developers to ingest massive quantities of data for use in training ML models with services such as Amazon SageMaker. AWS also makes it easy to host trained models as endpoints for making inferences against real-time data or deploying these models to the edge for local inferencing.

There are three kinds of software tools you will need to create and operate a purposeful, intelligent workload at the edge. Next, we will define each tool by its general capabilities and also the specific implementation of the tool we are using, provided by AWS, to build the project in this book. There is always more complexity to any **information technology (IT)**, but for our purposes, these are the three main kinds of tools this book will focus on in order to deliver intelligence to the edge.

Edge runtime

The first tool is a *runtime for orchestrating your edge software*. The runtime will execute your code and process local events to and from your code. Ideally, this runtime is self-healing, meaning that if any service fails, it should automatically recover by using failovers or restarting the service. These local events can be hardware interrupts that trigger some code to be run, timed events to read inputs from an analog sensor, or translating digital commands to change the state of a connected actuator such as a switch.

The AWS service that is the star of this book is **IoT Greengrass**. This is the service that we will use for the first kind of tool: the runtime for orchestrating edge software. IoT Greengrass defines both a packaged runtime for orchestrating edge software solutions and a web service for managing fleets of edge deployments running on devices such as gateways. In 2020, AWS released a new major version of IoT Greengrass, version 2, that rearchitected the edge software package as an open source Java project under the Apache 2.0 license. With this version, developers got a new software development model for authoring and deploying linked components that lets them focus on building business applications instead of worrying about the infrastructure of a complex edge-to-cloud solution. We will dive into more details of IoT Greengrass and start building our first application with it in the next chapter.

The following diagram illustrates how IoT Greengrass plays a role both at the edge and in the cloud:

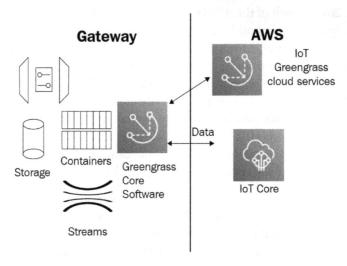

Figure 1.6 – Illustration of how IoT Greengrass plays a role both at the edge and in the cloud

ML

The second tool is an ML library and model. The library dictates how to read and consume the model and how local code can invoke the model, also called *making an inference*. The model is the output of a training job that packages up the intelligence into a simpler framework for translating new inputs into inferences. We will need a tool to train a new model from a set of data called the training set. Then, that trained model will be packaged up and deployed to our edge runtime tool. The edge solution will need the corresponding library and code that knows how to process new data against the model to yield an inference.

The implementation of our second tool, the ML library and model, is delivered by Amazon SageMaker. SageMaker is Amazon's suite of services for the ML developer. Included are services for preparing data for use in training, building models with built-in or custom algorithms, tuning models, and managing models as **application programming interface** (**API**) endpoints or deploying them wherever they need to run. You can even train a model without any prior experience, and SageMaker will analyze your dataset, select an algorithm, build a set of tuned models, and tell you which one is the best fit against your data.

For some AI use cases, such as forecasting numerical series and interpreting human handwriting as text, Amazon offers purpose-built services that have already solved the heavy lifting of training ML models. Please note that the teaching of data science is beyond the scope of this book. We will use popular ML frameworks and algorithms common for delivering outcomes in IoT solutions. We will also provide justification for frameworks and algorithms when we use them to give you some insight into how we arrived at choosing them. We will introduce the concept of deploying ML resources to the edge in *Chapter 4, Extending the Cloud to the Edge*, and dive deeper into ML workloads in *Chapter 7, Machine Learning Workloads at the Edge*, and *Chapter 9, Fleet Management at Scale*.

Communicating with the edge

The third tool is the methodology for communicating with the edge solution. This includes deploying the software and updates to the edge hardware and the mechanism for exchanging bi-directional data. Since the kinds of edge solutions that this book covers are those that interoperate with the cloud, a means of transmitting data and commands between the edge and the cloud is needed. This could be any number of **Open Systems Interconnection (OSI)** model Layer 7 protocols, with common examples in IoT being **HyperText Transfer Protocol (HTTP)**, **Message Queuing Telemetry Transport (MQTT)**, and **Constrained Application Protocol (CoAP)**.

The AWS IoT suite of services fits the needs here and acts as a bridge between the IoT Greengrass solution running at the edge and the ML capabilities we will use in Amazon SageMaker. IoT Core is a service that provides a scalable device gateway and message broker for both HTTP and MQTT protocols. It will handle the cloud connectivity, authentication and authorization, and routing of messages between the edge and the cloud. **IoT Device Management** is a service for operating fleets of devices at scale. It will help us define logical groupings of edge devices that will run the same software solutions and deploy updates to our fleet. Most chapters in this book will rely on tools such as these, and there is a focus on the scale of fleet management in *Chapter 8, DevOps and MLOps for the Edge*, and *Chapter 9, Fleet Management at Scale*.

With these tools in mind, we will next explore the markets, personas, and use cases where edge-to-cloud solutions with ML capabilities are driving the most demand.

Demand for smart home and industrial IoT

Market trends and analysis point to steep growth in the IoT industry, particularly in the industrial IoT segment. The 2020 *Mordor Intelligence* report *Smart Homes Market – Growth, Trends, COVID-19 Impact, and Forecasts (2021-2026)* projects the smart home market to grow from **$79 billion US Dollars (USD)** in 2020 to reach $313 billion by 2026. Similarly, the 2019 *Grand View Research* report *Industrial Internet Of Things Market Size, Share & Trends Analysis Report By Component (Solution, Services, Platform) By End Use (Manufacturing, Logistics & Transport), By Region, And Segment Forecasts, 2019-2025* projects the industrial IoT market to grow from $214 billion in 2018 to reach $949 billion by 2025. In both studies, the estimated **compound annual growth rate (CAGR)** is approximately 25-30%. That means there are big opportunities for new products, solutions, and services to find success with businesses and end consumers.

You can see a depiction of market forecasts for smart home and industrial IoT here:

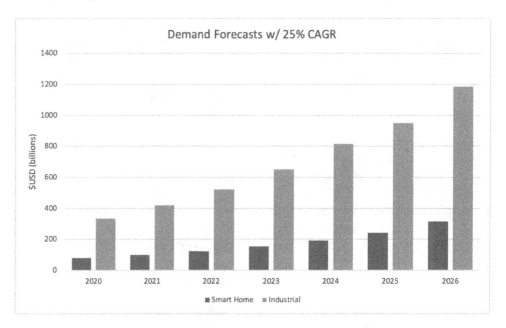

Figure 1.7 – Market forecasts for smart home and industrial IoT

It's important to keep in mind that forecasts are just that: forecasts. The only way those forecasts become reality is if inventors and problem solvers such as you and I get excited and make stuff! The key to understanding the future of smart home and industrial IoT solutions is how they are influenced by the value propositions of complete edge-to-cloud patterns and local ML inferencing. We can reflect on the key benefits of bringing ML to the edge to see how solutions in these markets are ripe for innovation.

Smart home use cases

In smart home solutions, the standard for functionality is oriented around environmental monitoring (temperature/electrical consumption/luminescence), automating state changes (turn this on in the morning and off at night), and introducing convenience where it was not previously possible (turn on the air conditioning when you are on your way home).

The primary persona using the product is the end consumer who lives in the residence where the solution is deployed. Secondary personas are guests of the owner, pets, public utilities, and home security service providers. At the product design level, the chief stakeholders are the IoT architect, security engineer, device manufacturer, and data scientist. Smart home products have been exploring and enjoying critical success when tapping into the power of AI and ML hosted in the cloud.

Here are three ways that deploying ML capabilities to the edge can benefit smart home use cases:

- **Voice-assisted interfaces**: Smart voice assistants such as Amazon's Alexa rely on the cloud to perform speech-to-text routines in order to process commands and generate audio responses. Running speech recognition models at the edge can help keep some common commands available for consumers even when the network is unavailable. Training models for recognition of who is speaking and incorporating that in responses increases the personalization factor and could make these voice assistants feel even more believable.

- **Home security**: Recognizing a breach of security has traditionally relied on binary sensors such as passive infrared for motion or magnetic proximity to detect open doors and windows. This simple mechanism can lead to false positives and undetected real security events. The next level of smart security will require complex event detection that analyzes multivariate inputs and confidence scores from trained models. Local models can evaluate whether the consumer is home or away automatically, and the solution can use that to calibrate sensitivity to events and escalate notifications of events. Video camera feeds are a classic example of a high data rate use case that becomes significantly cheaper to use with local processing for determining which clips to upload to the cloud for storage and further processing.

- **Sustainability and convenience**: Simple thermostats that maintain a temperature threshold are limited to recognizing when the threshold is breached and reacting by engaging a furnace or air conditioning system. Conventional smart home automation improves on this by reading weather forecasts, building a schedule profile of who is present in the home, and obeying rules for economical operation. ML can take us even further by analyzing a wider variety of inputs to determine via a recommendation engine how to achieve personal comfort targets most sustainably. For example, an ML model might identify and tell us that for your specific home, the most sustainable way to cool off in the evenings is to run the air conditioning in 5-minute bursts over 2 hours instead of frontloading for 30 minutes.

Industrial use cases

In industrial verticals such as manufacturing, power and utilities, and supply chain logistics, the common threads to innovating are creating profitable new business models and reducing the costs of existing business models. In order to innovate with the world of IT, these goals can be achieved through a better understanding of customer needs and the operational data generated by the business to test a new hypothesis. That understanding comes from using more of the existing data already collected and acquiring new streams of data needed to resolve hypotheses that lead to valuable new opportunities.

As per the 2015 *McKinsey Global Institute* report *Unlocking the potential of the Internet of Things*, only 1% of data collected by a business's IoT sensors is examined. The challenge to using the data is making it accessible to the systems and people that can get value from it. Data has little value when it is ingested at the edge but stored in an on-premises silo that can't afford to ship it to the cloud for analysis. This is where today's edge solutions can turn data into actionable insights with local compute and ML.

Here are three use cases for ML at the edge in industrial IoT settings:

- **Predictive maintenance**: Industrial businesses invest in and deploy expensive machinery to perform work. This machinery, such as a sheet metal press, **computer numerical control** (**CNC**) router, or an excavator, only performs optimally for so many duty cycles before a maintenance operation is needed or, worse, before they experience a failure while on the job. The need to keep machinery in top condition while minimizing downtime and expenses on unnecessary maintenance is a leading use case for industrial IoT and edge solutions. Training models and deploying them at the edge for predictive maintenance detection not only saves businesses from expensive downtime events but builds on the benefits of local ML by ensuring smooth operations in remote environments without high-speed or consistent network access.

- **Safety and security**: The physical safety and security of employees should be the top concern for any business. Safety first, as it goes. ML-powered edge solutions raise the bar on workplace safety with applications such as **computer vision (CV)** models to detect when an employee is about to enter a hazardous environment without the required safety equipment, such as a hard hat or safety vest. Similar solutions can also be used to detect when unauthorized personnel are entering (or trying to enter) a restricted area. When it comes to human safety, latency and availability are paramount, so running a fully functional solution at the edge means bringing the ML capabilities with it.

- **Quality assurance**: When a human operator is inspecting a component or finished product on a manufacturing line, they know which aspects of quality to inspect based on a trained reference (every batch of cookies should taste like this cookie), comparison to a specification (thickness, sheen, strength of aluminum foil), or human perception (do these two blocks of wood have reasonably similar wood grain to be used together?). ML innovates how manufacturers, for example, can capture the intuition of human quality inspectors to increase the scale and precision of their operations. With sensors such as cameras and CV models deployed to the manufacturing environment, it is feasible to inspect every component or final product (instead of an arbitrary sample) with a statistically consistent evaluation applied every time. This also brings a benefit to **quality assurance (QA)** teams by shifting the focus to inspecting solution performance instead of working on highly repetitive tasks dependent upon rapid subjective analysis. In other words, I'd rather QA a sample of 10,000 items passing inspection from an ML solution instead of every one of those 10,000 items. Running such a solution at the edge delivers on the key benefits of reducing overall data sent to the cloud and minimizing latency for the solution to produce results.

These use cases across smart homes and industry highlight the benefits that can be achieved with ML-powered edge solutions. Lofty forecasts on market growth in IoT are more likely to become reality if there are more developers out there bringing innovative new edge solutions to life! Let's review the smart home solution (and gratis product idea for someone out there to build) that will drive the hands-on material throughout this book.

Setting the scene: A modern smart home solution

The solution you will construct over the chapters of this book is one that models a gateway device for a modern smart home solution. That means we will use the context of a smart home hub for gathering sensor data, analyzing and processing that data, and controlling local devices as functions of detected events, schedules, and user commands. We selected the smart home context as the basis of our solution throughout the book because it is simple to understand and we anticipate many of our readers have read about or personally interacted with smart home controllers. That enables us to use the context of the smart home as a trope to rapidly move through the hands-on chapters and get to the good stuff. If your goals for applying the skills learned in this book reach into other domains, such as industrial IoT, worry not; the technologies and patterns used in this book are applicable beyond the smart home context.

Now, it's time to put on your imagination hat while we dive deeper into the scenario driving our new smart home product! Imagine you are an employee of *Home Base Solutions*, a company that specializes in bringing new smart home devices to market. *Home Base Solutions* delivers best-in-class features for customers outfitting their home with smart products for the first time or for experienced customers looking for better service by replacing an older smart home system.

For the next holiday season, *Home Base Solutions* wants to release a new smart home hub that offers customers something they haven't seen before: a product that includes sensors for monitoring the health of their existing large appliances (such as a furnace or dishwasher) and uses ML to recommend to owners when maintenance is needed. A maintenance recommendation is served when the ML model detects an anomaly in the data from the attached appliance monitoring kit. This functionality must continue to work even when the public internet connection is down or congested, so the ML component cannot operate exclusively on a server in the cloud.

You can see an illustration of the *Home Base Solutions* appliance monitoring kit here:

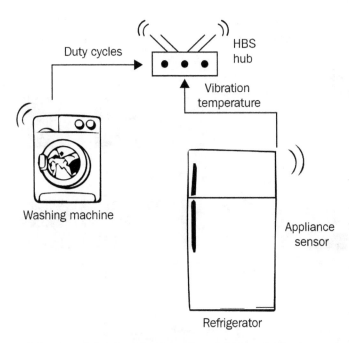

Figure 1.8 – Whiteboard sketch of the Home Base Solutions appliance monitoring kit

Your role in the company is the IoT architect, meaning you are responsible for designing the software solution that describes the E2E, edge-to-cloud model of data acquisition, ingestion, storage, analysis, and insight. It is up to you to design how to deliver upon the company's vision to incorporate ML technologies locally in the hub product such that there is no hard dependency on any remote service for continuous operation.

Being the architect also means you are responsible for selecting tools and designing how to operate a production fleet of these devices so that the customer service and fleet operations teams can manage customer devices at scale. You are not required to be a **subject-matter expert** (**SME**) on ML—that's where your team's data scientist will step in—but you should design an architecture that is compatible with feeding data to power ML training jobs and running built models on the hub device.

After spending a few weeks researching available software technologies, tools, solution vendors, and cloud services vendors, you decide to try out the following architecture using AWS:

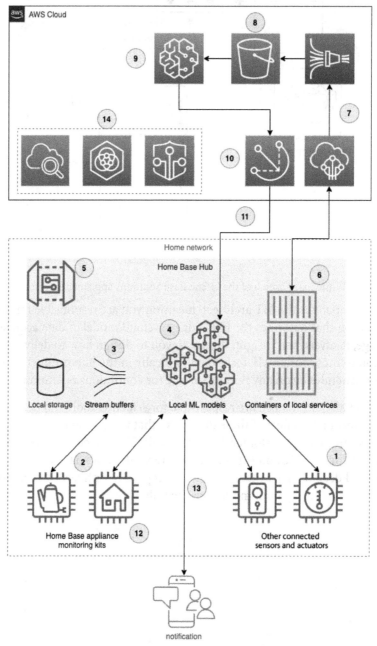

Figure 1.9 – Solution architecture diagram for appliance monitoring kit

It's just a little bit complicated, right? Don't worry if none of this makes sense yet. We will spend the rest of the book's chapters going into depth on these tools, introducing them at the beginner level, the patterns to use in production, and how to combine them to deliver outcomes. Each chapter will focus on one component or sub-section, and over time, we will work toward this total concept. The following is a breakdown of the individual components and their relationships:

1. Sensors and actuators controlled by the smart home hub, such as lights and their on/off switches or dimmers. These will be emulated with the Raspberry Pi Sense HAT, your own hardware modules compatible with Raspberry Pi, or via software components.

2. *Home Base Solutions* innovative home appliance monitoring kits. The streaming data from these kits will be implemented in this book as software components.

3. Stream buffer running on the smart home hub used to process home appliance runtime data.

4. ML models (one per home appliance) stored on the hub to invoke against incoming telemetry streamed from the home appliance monitoring kits.

5. This is all through an edge software solution running on the *Home Base Solutions* hub device and its components that are deployed and run by IoT Greengrass Core software.

6. Components running inside the IoT Greengrass edge solution exchange messages with the AWS cloud via MQTT messages and the AWS IoT Core service.

7. The rules engine of AWS IoT Core enables light **extract-transform-load** (**ETL**) operations and forwarding of messages throughout the cloud side of the solution.

8. Home appliance monitoring data is stored in Amazon **Simple Storage Service** (**S3**).

9. Amazon SageMaker uses appliance monitoring data as inputs for training and retraining ML models.

10. The cloud service of IoT Greengrass, using native features of IoT Device Management such as groups and jobs, deploys code and resources down to the fleet of smart home hubs.

11. Trained ML models are deployed as resources back to the edge for local inferences.

12. A local feedback mechanism, such as a **light-emitting diode** (**LED**) or speaker, signals to customers that an anomaly has been detected and that a maintenance activity is suggested.

13. A network feedback mechanism, such as a push notification to a mobile application, signals to customers that a maintenance activity is suggested and can provide further context about the anomalous event.

14. Customer support and fleet operations teams use a suite of tools such as Fleet Hub, Device Defender, and CloudWatch for monitoring the health of customer devices.

By the end of this book, you will have built this entire solution and have the skills needed to apply a similar solution to your own business needs. The patterns and overall shape of the architecture stay consistent. The implementation details of specific devices, networks, and outcomes needed are what vary from project to project.

Hands-on prerequisites

In order to follow along with the hands-on portions of this book, you will need access to two computer systems.

> **Note**
>
> At the time of authoring, AWS IoT Greengrass v2 did not support Windows installation. The hands-on portions related to the edge solution are specific to Linux and do not run on Windows

System 1: The edge device

The first system will be your **edge device**, also known as a gateway, since it will act as the proxy for one or more devices and the cloud component of the solution. In IoT Greengrass terminology, this is called a **Greengrass core**. This system must be a computer running a Linux operating system, or a **virtual machine** (**VM**) of a Linux system. The runtime software for AWS IoT Greengrass **version 2** (**v2**) has a dependency on Linux at the time of this writing. The recommendation for this book is to use a Raspberry Pi (hardware version 3B or later) running the latest version of Raspberry Pi OS. Suitable alternatives include a Linux laptop/desktop, a virtualization product such as VirtualBox running a Linux image, or a cloud-hosted Linux instance such as Amazon **Elastic Compute Cloud** (**EC2**), Azure Virtual Machines, or DigitalOcean Droplets.

The Raspberry Pi is preferred because it provides the easiest way to interoperate with physical sensors and actuators. After all, we are building an IoT project! That being said, we will provide code samples to emulate the functionality of sensors and actuators for our readers who are using virtual environments to complete the hands-on sections. The recommended expansion kit to cover use cases for sensors and actuators is the **Raspberry Pi Sense HAT**. There are many kits out there of expansion boards and modules compatible with the Raspberry Pi. The use cases in this book could be accomplished or modified as necessary to fit what you have, though we will not cover alternatives beyond the software samples provided.

You can see a visual representation of the Raspberry Pi 3B with a Sense HAT expansion board here:

Figure 1.10 – Raspberry Pi 3B with Sense HAT expansion board

In order to keep the **bill of materials** (**BOM**) low for the solution, we are setting the border of edge communications at the gateway device itself. This means that there are no devices wirelessly communicating with the gateway in this book's solution, although a real-world implementation for the smart home product would likely use some kind of wireless communication.

If you are using the recommended components outlined in this section, you will have access to an array of sensors, buttons, and feedback mechanisms that emulate interoperation between the smart home gateway device and the connected devices installed around the home. In that sense, the communication between devices and the smart home hub becomes an implementation detail that is orthogonal to the software design patterns showcased here.

System 2: Command and control (C2)

The second system will be your C2 environment. This system can be a Windows-, Mac-, or Unix-based operating system from which you will install and use the **AWS Command Line Interface** (**AWS CLI**) to configure, update, and manage your fleet of edge solutions. IoT Greengrass supports a local development life cycle, so we will use the edge device directly (or via **Secure Shell** (**SSH**) from the second system) to get started, and then in later chapters move exclusively to the C2 system for remote operation.

Here is a simple list of requirements:

- An AWS account
- A user in the AWS account with administrator permissions
- First system (edge device):

- Linux-based operating system such as Raspberry Pi OS or Ubuntu 18.x

  - Recommended: Raspberry Pi (hardware revision 3B or later)

  - Must be of architecture Armv7l, Armv8 (AArch64), or x86_64

- 1 **gigahertz (GHz) central processing unit (CPU)**

- 512 **megabytes (MB)** disk space

- 128 MB **random-access memory (RAM)**

- Keyboard and display (or SSH access to this system)

- A network connection that can reach the public internet on **Transmission Control Protocol (TCP)** ports 80, 443, and 8883

- sudo access for installing and upgrading packages via package manager

- (optional) Raspberry Pi Sense HAT or equivalent expansion modules for sensors and actuators

- Second system (C2 system):

  - Windows-, Mac-, or Unix-based operating system

  - Keyboard and display

  - Python 3.7+ installed

  - AWS CLI v2.2+ installed

  - A network connection that can reach the public internet on TCP ports 80 and 443

> **Note**
>
> If you are creating a new AWS account for this project, you will also need a credit card to complete the signup process. It is recommended to use a new developer account or sandbox account if provisioned by your company's AWS administrator. It is not recommended to experiment with new projects in any account running production services.

All of this is an exhaustive way of saying: if you have a laptop and a Raspberry Pi, you are likely ready to proceed! If you just have a laptop, you can still complete all of the hands-on exercises with a local VM at no additional cost.

> **Note**
>
> Installation instructions for Python and the AWS CLI vary per operating system. Setup for these tools is not covered in this book. See `https://www.python.org` and `https://aws.amazon.com/cli/` for installation and configuration.

Summary

You should now have a working definition of the edge of computing topology and the components of edge solutions such as sensors, actuators, and compute capability. You should be able to identify the value proposition of ML technology for smart home and industrial use cases running at the edge. We created an imaginary company, scoped a new product launch, and described the overall architecture of the solution you will deliver throughout the rest of this book.

In the next chapter, you will take your first steps toward developing the edge solution by learning how to orchestrate code on your edge device with AWS IoT Greengrass. If the prerequisites of your two hands-on systems are ready to go and your AWS account is set up, you are ready to go!

Knowledge check

Before moving on to the next chapter, test your knowledge by answering these questions. The answers can be found at the end of the book:

1. What's the difference between a cyber-physical solution and an edge solution?

2. At the time it was invented, the automobile was a self-contained mechanical entity, not a cyber-physical solution or an edge solution. At some point in the evolution of the automobile, it started meeting the definition of a cyber-physical solution, and then again meeting the definition of an edge solution. What are the characteristics of automobiles we can find today that meet our definition of an edge solution?

3. Has the telephone always been a cyber-physical solution? Why or why not?

4. What are the common components of an edge solution?

5. What are the three primary types of tools needed to deliver intelligence workloads at the edge?

6. What are the four key benefits in edge-to-cloud workloads that can be achieved with ML models running at the edge?

7. Who is the primary persona at the heart of any smart home solution?

8. Can you identify one more use case for the smart home vertical that ties in with one more of the key benefits for ML-powered edge solutions?

9. Who is the primary persona at the heart of any industrial solution?

10. Can you identify one more use case for any industrial vertical that ties in with one more of the key benefits of ML-powered edge solutions?

11. Is the IoT architect of an ML-powered edge solution typically responsible for the performance accuracy (for example, confidence scores for a prediction) of the models deployed? Why or why not?

References

Take a look at the following resources for additional information on the concepts discussed in this chapter:

- *Erich Gamma, Richard Helm, Ralph Johnson,* and *John Vlissides.* 1995. *Design Patterns: Elements of Reusable Object-Oriented Software. Addison-Wesley Longman Publishing Co., Inc., USA*:

- *Smart Homes Market – Growth, Trends, COVID-19 Impact, and Forecasts (2021-2026)*:

  https://www.mordorintelligence.com/industry-reports/global-smart-homes-market-industry

- *Industrial Internet Of Things Market Size, Share & Trends Analysis Report By Component, (Solution, Services, Platform), By End Use (Manufacturing, Logistics and Transport), By Region, And Segment Forecasts, 2019-2025*:

  https://www.grandviewresearch.com/industry-analysis/industrial-internet-of-things-iiot-market

- *Unlocking the potential of the Internet of Things*:

  https://www.mckinsey.com/business-functions/mckinsey-digital/our-insights/the-internet-of-things-the-value-of-digitizing-the-physical-world

- *The 17 Goals*, UN website:

  https://sdgs.un.org/goals

Section 2: Building Blocks

This section will work hands-on with AWS technologies such as IoT Greengrass and Amazon SageMaker to create a solution that delivers the power of machine learning models on local device data streams.

This section comprises the following chapters:

- *Chapter 2, Foundations of Edge Workloads*
- *Chapter 3, Building the Edge*
- *Chapter 4, Extending the Cloud to the Edge*
- *Chapter 5, Ingesting and Streaming Data from the Edge*
- *Chapter 6, Processing and Consuming Data on the Cloud*
- *Chapter 7, Machine Learning Workloads at the Edge*

2
Foundations of Edge Workloads

This chapter will explore the next level of detail regarding **edge workloads** and your first hands-on activity. You will learn how **AWS IoT Greengrass** meets the needs for designing and delivering modern edge ML solutions. You will learn how to prepare your edge device to work with AWS by deploying a tool that checks your device for compatible requirements. Additionally, you will learn how to install the IoT Greengrass core software and deploy your first IoT Greengrass core device. You will learn about the structure of components, examine the fundamental unit of software in IoT Greengrass, and write your first edge workload component.

By the end of this chapter, you should start to feel comfortable with the basics of IoT Greengrass and its local development life cycle.

In this chapter, we are going to cover the following main topics:

- The anatomy of an edge ML solution
- IoT Greengrass for the win
- Checking compatibility with IoT Device Tester
- Installing IoT Greengrass
- Your first edge component

Technical requirements

The technical requirements for this chapter are the same as those described in the Hands-on prerequisites section in *Chapter 1, Introduction to the Data-Driven Edge with Machine Learning*. Please refer to the full requirements mentioned in that chapter. As a reminder, you will need the following:

- A Linux-based system to deploy the IoT Greengrass software. A Raspberry Pi 3B, or later, is recommended. The installation instructions are similar to other Linux-based systems. Please refer to the following GitHub repository for further guidance when the hands-on steps differ for systems other than a Raspberry Pi.

- A system to install and use the AWS **Command-Line Interface (CLI)**, enabling access to the AWS Management Console website (typically, your PC/laptop).

You can access this chapter's technical resources from the GitHub repository, under the `chapter2` folder, at `https://github.com/PacktPublishing/Intelligent-Workloads-at-the-Edge/tree/main/chapter2`.

The anatomy of an edge ML solution

The previous chapter introduced the concept of an edge solution along with the three key kinds of tools that define an edge solution with ML applications. This chapter provides more detail regarding the layers of an edge solution. The three layers addressed in this section are as follows:

- The **business logic layer** includes the customized code that dictates the solution's behavior.

- The **physical interface layer** connects your solution to the analog world with sensors and actuators.

- The **network interface layer** connects your solution to other digital entities in the wider network.

Learning more about these layers is important because they will inform how you, as the IoT architect, make trade-offs when designing your edge ML solution. First, we'll start by defining the business logic layer.

Designing code for business logic

The business logic layer is where all the code of your edge solution lives. This code can take many shapes, such as *precompiled binaries* (such as a C program), *shell scripts, code evaluated by a runtime* (such as a Java or Python program), and *ML models*. Additionally, code can be organized in a few different ways such as shipping everything into a monolithic application, splitting up code into services or libraries, or bundling code to run in a container. All of these options come with implications for architecting and shipping an edge ML solution, such as security, cost, consistency, productivity, and durability. Some of the challenges of delivering code for the business logic layer are as follows:

- Writing and testing code that will run on your edge hardware platforms. For example, writing code that will work on variations of hardware platforms as incremental new versions are rolled out. You will want to minimize the number of forks of code you maintain that meet the needs of all your hardware platforms.

- Designing a robust edge solution that encompasses many features. For example, bundling features to process new sensor data, analyze data, and communicate with web services that do not create conflicts with common dependencies or local resources.

- Writing code with a team of people all working on an aggregate edge solution. For example, a monolithic application with many contributors can require each author to fully know the solution to make an incremental change.

To address the challenges of writing your business layer logic, the best practice for shipping code to the edge is to use **isolated services** where practical.

Isolated services

On your *Home Base Solutions hub device* (the fictional product we are creating from the story in *Chapter 1, Introduction to the Data-Driven Edge with Machine Learning*), code will be deployed and run as isolated services. In this context, a **service** is a self-contained unit of business logic that is either invoked by another entity to perform a task or performs a task on its own. **Isolation** means that the service will bundle with it the code, resources, and dependencies it needs for its operation. For example, a service you will create in *Chapter 7, Machine Learning Workloads at the Edge*, will run code to read from a data source or collection of images, periodically compute inferences using a bundled ML model, then publish any inference results to a local stream or the cloud. This pattern of isolated services is selected for two reasons:

- The first reason is that a **service-oriented architecture** enables architects to design capabilities that are decoupled from one another. **Decoupling** means we use data structures, such as buffers, queues, and streams, to add a layer of abstraction between our services, reducing dependencies to allow services to run independently.

  You can deploy updates to individual services without touching other running services and, therefore, reduce the risk of impact on them. Decoupled, service-oriented architecture is a best practice for designing well-architected cloud solutions that are also a good fit for edge ML solutions where multiple services are simultaneously running and emphasize a need for reliability. For example, a service that interfaces a sensor writes new measurements to a data structure and nothing more; it has a single job and doesn't need to be aware of how the data is consumed by a later capability.

- The second reason is that the **isolation** of code enables developers to focus on what that code does instead of where the code is going or how dependencies are managed at the destination. By using principles of isolation to bundle runtime dependencies and resources with code, we get stronger consistency that code will work deterministically wherever it is deployed. Developers free up the effort required for dependency management and have more confidence that the code will work the same way on the edge platform, which is likely different than their development environment. That's not to say an edge solution developer won't need to test the behavior of their code against physical interfaces such as sensors and actuators. However, it does mean that development teams can deliver self-contained services that work independently regardless of the rest of the services deployed in the aggregate edge solution.

Examples of isolation include **Python** virtual environments, which explicitly specify a Python runtime version and package, and **Docker Engine**, which uses containers to bundle dependencies, resources, and achieve process isolation on the host. The following diagram illustrates the separation of concerns achieved with isolated services:

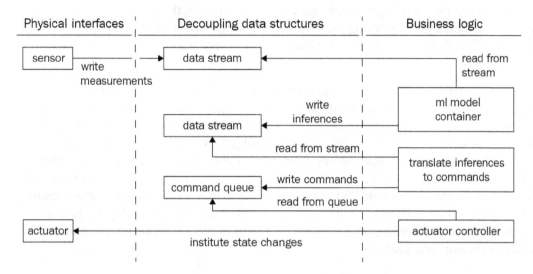

Figure 2.1 – An edge solution using decoupled, isolated services

In tandem, the patterns of isolation and services offer compelling benefits for edge ML solutions. Of course, every decision in development comes with trade-offs. The solution would be simpler if deployed as a singular monolith of code and faster to derive a minimum viable product. We opt for more architectural complexity because this leads to better resiliency and a scaling up of the solution over time. We lean on strong patterns and good tooling to balance that complexity.

IoT Greengrass is designed with this pattern in mind. Later, in this chapter and throughout the book, you will learn how to use this pattern with IoT Greengrass to develop well-architected edge ML solutions.

Physical interfaces

A cyber-physical solution is defined by the use of physical interfaces to interact with the analog world. These interfaces come in two classifications: *sensors for taking measurements from the analog world* and *actuators for exerting change back upon it.* Some machines do both, such as a refrigerator that senses internal temperature and activates its compressor to cycle a refrigerant through coils. In these cases, the aggregation of sensing and actuating is logical, meaning the sensor and actuator have a relationship but are functionally independent and are coordinated via a mechanism such as a switch, circuit, or microcontroller.

Sensors that perform analog-to-digital conversion do so by sampling the voltage from an electrical signal and converting it into a digital value. These digital values are interpreted by code to derive data such as temperature, light, and pressure. Actuators convert digital signals into analog actions, typically, by manipulating the voltage going to a switch or circuit. A command to engage a motor is interpreted as raising a voltage to the level that activates the circuit. Diving deeper into the electrical engineering of physical interfaces is beyond the scope of this book. Please refer to the *References* section for recommendations on deeper dives on that topic. The following diagram shows a simple analog example of a refrigerator and the relationship between the thermostat (sensor), switch (controller), and compressor (actuator):

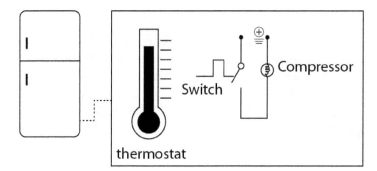

Figure 2.2 – An analog controller with a sensor and an actuator

It's important to understand the patterns of input and output delivered by a cyber-physical solution and the relationship to a higher-level outcome delivered through an edge ML solution. Throughout the project delivered in this book, you will gain hands-on experience applying these patterns. Some of the services of the Home Base Solutions hub device will serve as interfaces to the physical layer, providing new measurements from the sensors and converting the commands to change the state of local devices. If you are working with a physical edge device, such as the Raspberry Pi, you will get some experience of using code to interact with the physical interfaces of that device.

Network interfaces

The third layer to introduce for our edge solution anatomy is the *network interface*. A differentiator between our definitions of cyber-physical solutions and edge solutions is that an edge solution will, at some point, interact with another entity over a network. For example, the design of our new appliance monitoring kit for Home Base Solutions uses wireless communication between the monitoring kit and the hub device. There is no physical connection between the two for the purposes of analog-to-digital signal conversion from the monitor's sensors.

Similarly, the hub device will also exchange messages with a cloud service to store telemetry to use in the training of the ML model, to deploy new resources to the device, and to alert customers of recognized events. The following diagram illustrates the flow of messages and the relationships between a **Sensor**, **Actuator**, hub device (**Gateway**), and the cloud service:

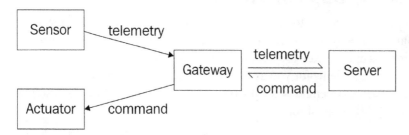

Figure 2.3 – An edge device exchanging messages with local sensors, actuators, and the cloud

Wireless communications are common in IoT solutions, and specific implementations enable connectivity at a wide range of distances. Each specification and implementation makes trade-offs for range, data transmission rate, hardware cost, and energy consumption. Short-range radio specifications such as **Zigbee** (IEEE 802.15.4), **Bluetooth** (IEEE 802.15.1), and **WiFi** (IEEE 802.11) are suitable for bridging devices within personal and local area networks. Long-range radio specifications such as conventional cellular networks (for example, **GSM**, **CDMA**, and **LTE**) and **low-power wide-area networks** (**LPWANs**) such as **LoRaWAN** and **NB-IoT** deliver connectivity options for devices deployed (either static or roaming) across a particular campus, city, or region.

Wired communications are still common for connecting edge devices such as TVs, game consoles, and PCs to home network solutions such as switches and routers over **Ethernet**. Wired connectivity is less common in smart home products due to the limited number of Ethernet ports on home networking routers (typically, there are only 1–8 ports), the restrictions regarding where the device can be placed, and the burden of adding wires throughout the home.

For example, the Home Base Solutions appliance monitoring kit would likely use a Zigbee, or equivalent implementation, and a battery to balance energy consumption with the anticipated data rate. If the kit required a power supply from a nearby outlet, Wi-Fi becomes more of an option; however, it would limit the overall product utility as the placement of the kinds of appliances to monitor don't always have a spare outlet. Additionally, it wouldn't make sense to use Ethernet to connect the kits to the hub directly since customers likely wouldn't find all of the extra wires running throughout the home appealing. The hub device that communicates with the kit could use Ethernet or Wi-Fi to bridge to the customer's local network to gain access to the public internet.

Now that you have a better understanding of the three layers of an edge solution, let's evaluate the selected edge runtime solution and how it implements each layer.

IoT Greengrass for the win

The most important question to answer in a book about using IoT Greengrass to deliver edge ML solutions is *why IoT Greengrass?* When evaluating the unique challenges of edge ML solutions and the key tools required to deliver them, you want to select tools that solve as many problems for you as possible while staying out of your way in terms of being productive. IoT Greengrass is a purpose-built tool with IoT and ML solutions at the forefront of the value proposition.

IoT Greengrass is prescriptive in how it solves the problem of the *undifferentiated heavy lifting* of common requirements while remaining non-prescriptive in how you implement your business logic. This means that the out-of-box experience yields many capabilities for rapid iteration without being obstructive about how you use them to reach your end goals. The following is a list of some of the capabilities that IoT Greengrass has to offer:

- **Security at the edge**: IoT Greengrass is installed with root permissions and uses operating system user permissions to protect the code and resources deployed at the edge from tampering.

- **Security to the cloud**: IoT Greengrass uses mutual **transport layer security** (**TLS**) with public key infrastructure to exchange messages between the edge and the cloud. Resources are fetched during deployments using HTTPS and AWS Signature Version 4 to verify the identity of the requester and protect data in transit.

- **Runtime orchestration**: Developers can design applications however they prefer (with monoliths, services, or containers) and deploy them to the edge with ease. IoT Greengrass provides hooks for smartly integrating with component life cycle events, or developers can ignore them and simply bootstrap applications with a single command. Individual components can be added or updated without interrupting other running services. A dependency tree allows developers to abstract out the installation of libraries and configuration activities to decouple from code artifacts.

- **Logging and monitoring**: By default, IoT Greengrass creates logs for each component and allows developers to indicate which log files should be synchronized to the cloud for operational purposes. Additionally, the cloud service keeps track of device health automatically, making it easier for team members to identify and respond to unhealthy devices.

- **Scaling up the fleet**: Deploying updates to one device is not much different than deploying updates to a fleet of devices. It is easy to define groups, classify similar devices together, and then push updates to groups of devices using a managed deployment service.

- **Native integrations**: AWS provides many components to deploy into IoT Greengrass solutions that augment the baseline functionality and also for integrating with other AWS services. A stream management component enables you to define, write to, and consume streams at the edge. A Docker application manager allows you to download Docker images from public repositories or private repositories in **Amazon Elastic Container Registry**. Pretrained and optimized ML models are available for tasks such as object detection and image classification with **Deep Learning Runtime** and **TensorFlow Lite**.

In playing your role as the Home Base Solutions architect with a solution to build, you could propose that the engineering team invest the time and resources to build out all this functionality and test that it is production-ready. However, the IoT Greengrass baseline services and optional add-ons are ready to accelerate the development life cycle and come vetted by AWS, where *security is the top priority*.

IoT Greengrass does not do everything for you. In fact, a fresh installation of IoT Greengrass doesn't do anything but wait for further instruction in the form of a deployment. Think of it as a blank canvas, paint, and brush. It's everything you need to get started, but you have to develop the solution that it runs. Let's review the operating model for IoT Greengrass, both at the edge and within the cloud.

Reviewing IoT Greengrass architecture

IoT Greengrass is both a managed service running on AWS and an edge runtime tool. The managed service is where your devices are defined individually and in groups. When you want to push a new deployment to the edge, you are, in fact, invoking an API in the managed service that is then responsible for communicating with the edge runtime to coordinate the delivery of that deployment. Here is a sequence diagram showing the process of you, as the developer, configuring a component and requesting that a device running the IoT Greengrass core software receive and run that component:

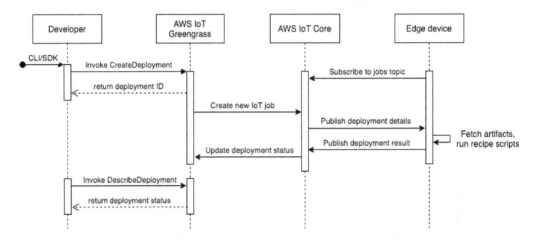

Figure 2.4 – Pushing a deployment through IoT Greengrass

The **component** is the fundamental unit of functionality that is deployed to a device running IoT Greengrass. Components are defined by a manifest file, called a **recipe**, which tells IoT Greengrass what the name, version, dependencies, and instructions are for that component. In addition to this, a component can define zero or more **artifacts** that are fetched during deployment. These artifacts can be binaries, source code, compiled code, archives, images, or data files; really any kind of file or resource that is stored on disk. Component recipes can define dependencies on other components that get resolved via a graph by the IoT Greengrass software.

During a deployment activity, one or more components are added or updated in that deployment. The component's artifacts are downloaded to the local device from the cloud; then, the component is started by way of evaluating life cycle instructions. A life cycle instruction could be something that happens during startup; it could be the main command to run, such as starting a Java application or something to do after the component has concluded running. Components might continue in the running state indefinitely or perform a task and exit. The following diagram provides an example of the component graph:

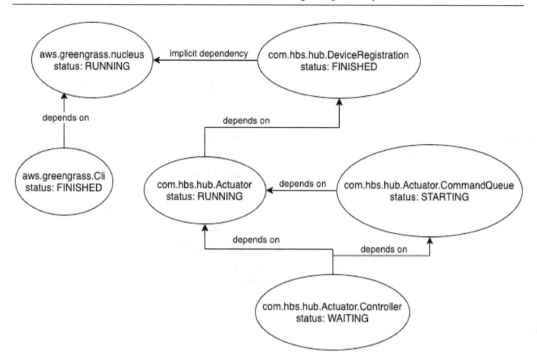

Figure 2.5 – An example graph of components showing the life cycle and dependencies

That's everything we need to cover before taking your first steps toward readying your edge device to run a solution with IoT Greengrass!

In the following sections, you will validate that your edge device is ready to run the IoT Greengrass software, install the software, and then author your first component. In addition to this, you will get a closer look at components and deployments through the hands-on activities to come.

Checking compatibility with IoT Device Tester

IoT Device Tester (**IDT**) is a software client provided by AWS to assess device readiness for use in AWS IoT solutions. It assists developers by running a series of qualification tests to validate whether the destination system is ready to run the IoT Greengrass core software. Additionally, it runs a series of tests to prove that edge capabilities are already present, such as establishing MQTT connections to AWS and running ML models locally. IDT works for one device you are testing locally or scales up to run customized test suites against any number of device groups, so long as they are accessible over the network.

In the context of your role as the IoT architect at Home Base Solutions, you should use IDT to prove that your target edge device platform (in this case, platform refers to hardware and software) is capable of running the runtime orchestration tool of choice, IoT Greengrass. This pattern of using tools to prove compatibility is a best practice over manually evaluating the target platform and/or assuming that a certain combination of listed requirements is met. For example, a potential device platform might advertise that its hardware requirements and operating system meet your needs, but it could be missing a critical library dependency that doesn't surface until later in the development life cycle. It is best to prove early that everything you need is present and accounted for.

> **Note**
>
> IDT does more than qualify hardware for running IoT Greengrass core software. The tool can also qualify hardware running FreeRTOS to validate that the device is capable of interoperating with AWS IoT Core. Developers can write their own custom tests and bundle them into suites to incorporate into your **software development life cycle (SDLC)**.

The following steps will enable you to prepare your Raspberry Pi device for use as the edge system (that is, the Home Base Solutions hub device in our fictional project) and configure your AWS account before finally running the IDT software. Optionally, you can skip the *Booting the Raspberry Pi* and *Configuring the AWS account and permissions* sections if you already have a device and AWS account configured for use. If you are using a different platform for your edge device, you just need to ensure you can reach the device over SSH with a system user that has root permissions from your command-and-control device.

Booting the Raspberry Pi

The following steps were run on a Raspberry Pi 3B with a clean installation of the May 2021 release of the Raspberry Pi OS. Please refer to `https://www.raspberrypi.org/software/` for the **Raspberry Pi Imager** tool. Use your command-and-control system to run the imaging tool to flash a Micro SD card with a fresh image of the Raspberry Pi OS. For this book's project, we recommended that you use a blank slate to avoid any unforeseen consequences of preexisting software and configuration changes. Here is a screenshot of the Raspberry Pi Imager tool and the image to select:

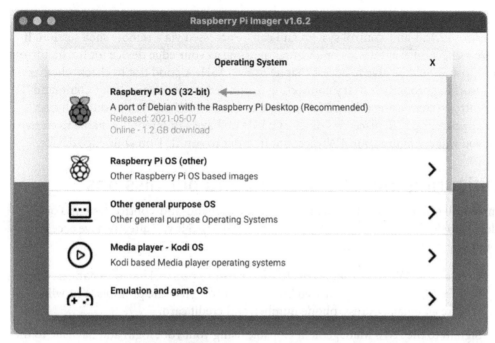

Figure 2.6 – The Raspberry Pi Imager tool and the image to select

The following is a list of steps to perform after flashing the Micro SD card with the Imager tool:

1. Insert the Micro SD card into the Raspberry Pi.

2. Turn on the Raspberry Pi by plugging in the power source.

3. Complete the first-time boot wizard. Update the default password, set your locale preferences, and connect to Wi-Fi (this is optional if you are using Ethernet).

4. Open the Terminal app and run `sudo apt-get update` and `sudo apt-get upgrade`.

5. Reboot the Pi.

6. Open the Terminal app and run the `hostname` command. Copy the value and make a note of it; for example, write it in a scratch file on your command-and-control system. On Raspberry Pi devices, the default is `raspberrypi`.

7. Open the Raspberry Pi preferences app and enable the SSH interface. This must be enabled for the IDT to access the device. Open **Preferences**, choose **Raspberry Pi Configuration**, choose **Interfaces**, and enable SSH.

At this milestone, your Raspberry Pi device is configured to join the same local network of your command-and-control system and can be accessed via a remote shell session. If you are using a different device or a virtual machine as your edge device for the hands-on project, you should be able to access that device via SSH. A good test to check whether this is working properly is to try connecting to your edge device from your command-and-control system using a Terminal application (or **PuTTY** on Windows), with `ssh pi@raspberrypi`. Replace `raspberrypi` if you have a different hostname from *step 6*. Next, you will configure your AWS account in order to run IDT on your edge device.

Configuring the AWS account and permissions

Complete all of the steps in this section on your command-and-control system. For readers who don't yet have an AWS account (skip to *step 5* if you already have access to an account), perform the following:

1. Create your AWS account. Navigate to `https://portal.aws.amazon.com/billing/signup` in your web browser and complete the prompts. You will require an email address, phone number, and credit card.

2. Sign in to the AWS Management Console using your root login and navigate to the **Identity and Access Management (IAM)** service. You can find this at `https://console.aws.amazon.com/iam/`.

3. Use the IAM service console to set up your administrative group and user account. It is a best practice to create a new user for yourself instead of continuing to use the root user login. You will use that new user to complete any later AWS steps using the AWS management console or AWS CLI:

    A. Create a new user called **Admin**. Select both the **Programmatic access** and **AWS Management Console access** types. Skip the sections for permissions and tags. Choose **Create user** in the **Review** section. In the confirmation section, make a note in your scratch file of the dedicated sign-in link for your account. Ensure that you download the CSV file with your AWS credentials, referred to as **Access Key** and **Secret Key**, which you will use to interact with the AWS CLI. Note that the CSV file also includes the user's password that is later used to log in to the AWS management console (instead of the current root user).

    B. Create a new user group called **Administrators**, add the **Admin** user to it, and attach to it the policy named `AdministratorAccess` (it is easier to use the filter field and type that in). This policy is managed by AWS to grant administrator-level permissions to users. The best practice is to relate permissions to groups and then assign users to groups to inherit permissions. This makes it easier to audit permissions and understand the kinds of access that users have from well-named groups.

4. Log out of the AWS management console.

> **Note**
>
> At this point, you should have access to an AWS account with an administrative user. Complete the following steps to set up the AWS CLI (please skip to *step 7* if you already have the AWS CLI configured with your admin user).

5. Install the AWS CLI. Platform-specific instructions can be found at `https://aws.amazon.com/cli/`. In this book, the AWS CLI steps will use AWS CLI v2.

6. Once installed, configure the AWS CLI and use the credentials you downloaded for the **Admin** user. This user will be stored as the *default* profile in your local AWS settings:

    A. In your Terminal or PowerShell application, run `aws configure`.

    B. When prompted for **AWS Access Key ID** and **AWS Secret Access Key**, use the values from the downloaded credentials file in *step 3A*. Do not use the password value here.

    C. When prompted for a default region, you can specify the AWS region you want to use by default for any commands. Note that some of the AWS CLI steps mentioned in this book require you to provide an explicit region in the command. at other times, the region is implicitly selected from this configuration step. In this book, the steps will use region **us-west-2** by default and remind readers to substitute when applicable.

    D. When prompted for a default output format, you can choose any of [`json`, `yaml`, `text`, `table`]. The author's preference is `json` and will be reflected in any examples of AWS CLI output that appear in the book.

    Next, you will use your `Admin` user to create a few more resources in preparation for the following sections to use the IDT and install the IoT Greengrass core software. These are permissions resources, similar to your `Admin` user, which will be used by IDT and IoT Greengrass software to interact with AWS on your behalf.

7. Log in to the AWS management console using your custom sign-in link from *step 3A*. Use the `Admin` username and password provided in the CSV file containing credentials.

8. Return to the IAM service console, which can be found at `https://console.aws.amazon.com/iam/`.

9.   Create a new user named `idtgg` (short for *IDT and Greengrass*) and select the **Programmatic access** type. This user will not require a password for management console access. Skip the permissions and tags sections. Make sure that you download the CSV file containing the credentials for this user as well.

10. Create a new policy. The first step is to define the permissions for the policy. Choose the **JSON** tab and paste the contents from this accompanying file into the book resources repository: `chapter2/policies/idtgg-policy.json`. Skip the tags section. In the review section, enter the name of `idt-gg-permissions`, enter the description of **permissions for IoT Device Tester and IoT Greengrass**, and choose **Create policy**.

11. Create a new user group with the name of **Provision-IDT-Greengrass**, select the `idt-gg-permissions` policy, and select the `idtgg` user. Choose **Create group**. You have now set up a new group, attached permissions, and assigned the programmatic access user that will serve as authentication and authorization for the IDT client and IoT Greengrass provisioning tool.

12. In your Terminal or PowerShell application, configure a new AWS CLI profile for this new user:

   A. Run `aws configure --profile idtgg`.

   B. When prompted for access keys and secret keys, use the new values from the credentials CSV file downloaded in *step 9*.

   C. When prompted for the default region, use the book's default of **us-west-2** or the AWS region you are using for all of the projects in this book.

That concludes all of the preparatory steps to configure your AWS account, permissions, and CLI. The next milestone is to install the IDT client and prepare it to test the Home Base Solutions prototype hub device.

Configuring IDT

You will run the IDT software from your command-and-control system, and the IDT will access the edge device system remotely through SSH to run the tests.

> **Note**
>
> The following steps reflect the configuration and use of IDT at the time of writing. If you get stuck, it might be that the latest version differs from the version that we used at the time of writing. You can check the AWS documentation for IDT for the latest guidance on installation, configuration, and use. Please refer to `https://docs.aws.amazon.com/greengrass/v2/developerguide/device-tester-for-greengrass-ug.html`.

Follow these steps to use IDT to verify the edge device system is ready to run IoT Greengrass. All of the following steps are completed on macOS using IoT Greengrass core software v2.4.0 and IDT v4.2.0 with test suite GGV2Q_2.0.1. For Windows, Linux, or later AWS software versions, please alter the commands and directories as needed:

1. On your command-and-control system, open a web browser and navigate to `https://docs.aws.amazon.com/greengrass/v2/developerguide/dev-test-versions.html`.

2. Underneath **Latest IDT version for AWS IoT Greengrass** and **IDT software downloads,** click on the link for the platform that matches your system. This will open a file download prompt. Save the archive to your local file system; for example, `C:\projects\idt` on Windows or `~/projects/idt` on macOS and Linux:

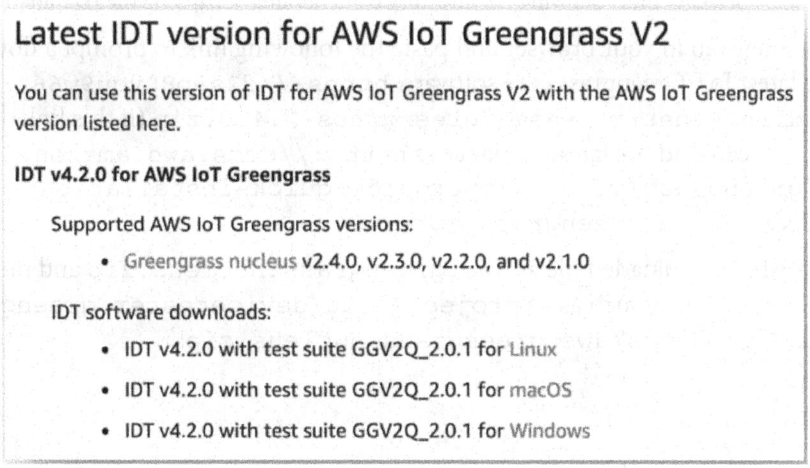

Latest IDT version for AWS IoT Greengrass V2

You can use this version of IDT for AWS IoT Greengrass V2 with the AWS IoT Greengrass version listed here.

IDT v4.2.0 for AWS IoT Greengrass

Supported AWS IoT Greengrass versions:

- Greengrass nucleus v2.4.0, v2.3.0, v2.2.0, and v2.1.0

IDT software downloads:

- IDT v4.2.0 with test suite GGV2Q_2.0.1 for Linux
- IDT v4.2.0 with test suite GGV2Q_2.0.1 for macOS
- IDT v4.2.0 with test suite GGV2Q_2.0.1 for Windows

Figure 2.7 – The AWS documentation website for downloading IDT; exact text and versions might differ

3. Unzip the archive contents in place in the directory. In a file explorer, double-click on the archive to extract them. If using Terminal, use a command such as `unzip devicetester_greengrass_v2_4.0.2_testsuite_1.1.0_mac.zip`. This is what the directory looks like on macOS:

Figure 2.8 – macOS Finder showing the directory contents after unzipping the IDT archive

4. Open a new tab in your browser and paste the following link to prompt a download of the latest IoT Greengrass core software: `https://d2s8p88vqu9w66.cloudfront.net/releases/greengrass-2.4.0.zip` (if this link doesn't work, you can find the latest guidance at `https://docs.aws.amazon.com/greengrass/v2/developerguide/quick-installation.html#download-greengrass-core-v2`).

5. Rename the downloaded file as `aws.greengrass.nucleus.zip` and move it to an IDT directory such as `~/projects/idt/devicetester_greengrass_v2_mac/products/aws.greengrass.nucleus.zip`:

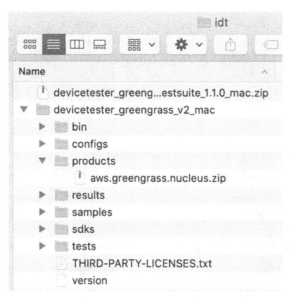

Figure 2.9 – The IoT Greengrass software in place to be used by IDT

6. Open a text file such as `~/projects/idt/devicetester_greengrass_v2_mac/configs/config.json` and update the following values:

    A. (Optionally) update `awsRegion` if you are not using the book's default of `us-west-2`.

    B. To use the `idtgg` profile that you configured earlier, set the value of `auth` as follows:

```
"auth": {
  "method": "file",
  "credentials": {
    "profile": "idtgg"
  }
}
```

7. Open a text file such as `~/projects/idt/devicetester_greengrass_v2_mac/configs/device.json` and update the following values:

   A. `"id": "pool1"`.

   B. `"sku": "hbshub"` (hbshub stands for Home Base Solutions hub).

   C. Underneath `"features"`, for the name-value pair named `"arch"`, set `"value": "armv7l"` (this is for Raspberry Pi devices; alternatively, you can choose the appropriate architecture for your device).

   D. Underneath `"features"`, for the remaining name-value pairs such as `"ml"`, `"docker"`, and `"streamManagement"`, set `"value": "no"`. For now, we will disable these tests because we have no immediate plans to use the tested features. Feel free to enable them if you'd like to evaluate your device's compatibility, although expect the tests to fail on a freshly imaged Raspberry Pi.

   E. Underneath `"devices"`, set `"id": "raspberrypi"` (or any device ID you prefer).

   F. Underneath `"connectivity"`, set the value of `"ip"` to the IP address of your edge device (for Raspberry Pi users, the value is the output of *step 6* from the *Booting the Raspberry Pi* section).

   G. Underneath `"auth"`, set `"method": "password"`.

   H. Underneath `"credentials"`, set the value of `"user"` to the username used to SSH to the edge device (typically, this will be `"pi"` for Raspberry Pi users).

   I. Underneath `"credentials"`, set the value of `"password"` to the password used to SSH to the edge device.

   J. Underneath `"credentials"`, delete the line for `"privKeyPath"`.

   K. Save the changes to this file. You can view a sample version of this file in the book's GitHub repository at `chapter2/policies/idt-device-sample.json`.

8. Open a text file such as `~/projects/idt/devicetester_greengrass_v2_mac/configs/userdata.json` and update the following values. Ensure that you specify absolute paths instead of relative paths:

   A. `"TempResourcesDirOnDevice": "/tmp/idt"`.

   B. `"InstallationDirRootOnDevice": "/greengrass"`.

   C. `"GreengrassNucleusZip": "Users/ryan/projects/idt/devicetester_greengrass_v2_mac/products/aws.greengrass.nucleus.zip"` (update this based on where you stored the `aws.greengrass.nucleus.zip` file in *step 5* of this section).

D. Save the changes to this file. You can view a sample version of this file in the book's GitHub repository at `chapter2/policies/idt-userdata-sample.json`.

9. Open an application such as Terminal on macOS/Linux or PowerShell on Windows.

10. Change your present working directory to where the IDT launcher is located:

   A. `~/projects/idt/devicetester_greengrass_v2_mac/bin` on macOS

   B. `~/projects/idt/devicetester_greengrass_v2_linux/bin` on Linux

   C. `C:\projects\idt\devicetester_greengrass_v2_win\bin` on Windows

11. Run the command to start IDT:

   A. `./devicetester_mac_x86-64 run-suite --userdata userdata.json` on macOS

   B. `./devicetester_linux_x86-64 run-suite --userdata userdata.json` on Linux

   C. `devicetester_win_x86-64.exe run-suite --userdata userdata.json` on Windows

Running IDT will start a local application that connects to your edge device over SSH and completes the series of tests. It will stop upon encountering the first failed test case or run until all test cases have passed. If you are running IDT against a fresh installation of your Raspberry Pi, as defined by previous steps, you should observe an output similar to the following:

```
========== Test Summary ==========
Execution Time:    3s
Tests Completed:   2
Tests Passed:      1
Tests Failed:      1
Tests Skipped:     0

----------------------------------

Test Groups:
    pretestvalidation:      PASSED
    coredependencies:       FAILED
```

```
------------------------------------
Failed Tests:
    Group Name: coredependencies
        Test Name: javaversion
            Reason: Encountered error while fetching java
version on the device: Failed to run Java version command
with error: Command '{java -version 2>&1 map[] 30s}'
exited with code 127. Error output: .
```

The step to install Java on the Raspberry Pi was intentionally left out in order to demonstrate how IDT identifies missing dependencies; apologies for the deception! If you ran the IDT test suite and passed all of the test cases, then you are ahead of schedule and can skip to the *Installing IoT Greengrass* section.

12. To fix this missing dependency, return to your Raspberry Pi interface and open the Terminal app.

13. Install Java on the Pi using sudo apt-get install default-jdk.

14. Return to your command-and-control system and run IDT again (repeat the command in *step 11*).

Your test suite should now pass the Java requirement test. If you encounter other failures, you will need to use the test report and logs in the idt/devicetester_ greengrass_v2_mac/results folder to triage and fix them. Some common missteps include missing AWS credentials, AWS credentials without sufficient permissions, and incorrect paths to resources defined in userdata.json. A fully passed suite of test cases looks like this:

```
========== Test Summary ==========
Execution Time:    22m59s
Tests Completed:   7
Tests Passed:      7
Tests Failed:      0
Tests Skipped:     0

------------------------------------
Test Groups:
    pretestvalidation:    PASSED
    coredependencies:     PASSED
    version:         PASSED
    component:         PASSED
    lambdadeployment:     PASSED
```

```
mqtt:              PASSED
------------------------------------
```

This concludes the introductory use of IDT to analyze and assist in preparing your devices to use IoT Greengrass. Here, the best practice is to use software tests, not just for your own code but to assess whether the edge device itself is ready to work with your solution. Lean on tools such as IDT that do the heavy lifting of proving that the device is ready to use and validate this for each new type of device enrolled or major solution version released. You should be able to configure IDT for your next project and qualify a new device or group of devices to run IoT Greengrass. In the next section, you will learn how to install IoT Greengrass on your device in order to configure your first edge component.

Installing IoT Greengrass

Now that you have used IDT to validate that your edge device is compatible with IoT Greengrass, the next milestone in this chapter is to install IoT Greengrass.

From your edge device (that is, the prototype Home Base Solutions hub), open the Terminal app, or use your command-and-control device to remotely access it using SSH:

1. Change the directory to your user's home directory: `cd ~/`.

2. Download the IoT Greengrass core software: `curl -s https://d2s8p88vqu9w66.cloudfront.net/releases/greengrass-nucleus-latest.zip > greengrass-nucleus-latest.zip`.

3. Unzip the archive: `unzip greengrass-nucleus-latest.zip -d greengrass && rm greengrass-nucleus-latest.zip`.

4. Your edge device requires AWS credentials in order to provision cloud resources on your behalf. You can use the same credentials that you created for the `idtgg` user in the previous *Configuring the AWS account and permissions* section:

    A. `export AWS_ACCESS_KEY_ID=AKIAIOSFODNN7EXAMPLE`

    B. `export AWS_SECRET_ACCESS_KEY=wJalrXUtnFEMI/K7MDENG/bPxRfiCYEXAMPLEKEY`

5. Install the IoT Greengrass core software using the following command. If you are using an AWS region other than `us-west-2`, update the value of the `--aws-region` argument. You can copy and paste this command from `chapter2/commands/provision-greengrass.sh`:

```
sudo -E java -Droot="/greengrass/v2" -Dlog.store=FILE \
  -jar ./greengrass/lib/Greengrass.jar \
```

```
    --aws-region us-west-2 \
    --thing-name hbshub001 \
    --thing-group-name hbshubprototypes \
    --tes-role-name GreengrassV2TokenExchangeRole \
    --tes-role-alias-name
GreengrassCoreTokenExchangeRoleAlias \
    --component-default-user ggc_user:ggc_group \
    --provision true \
    --setup-system-service true \
    --deploy-dev-tools true
```

6. That's it! The last few lines of output from this provisioning command should look like this:

```
Created device configuration
Successfully configured Nucleus with provisioned resource
details!
Creating a deployment for Greengrass first party
components to the thing group
Configured Nucleus to deploy aws.greengrass.Cli component
Successfully set up Nucleus as a system service
```

The installation of the IoT Greengrass core software and the provisioning of the initial resources is much smoother after validating compatibility with the IDT suite. Now your edge device has the first fundamental tool installed: the runtime orchestrator. Let's review the resources that have been created at the edge and in AWS from this provisioning step.

Reviewing what has been created so far

On your edge device, the IoT Greengrass software was installed in the /greengrass/v2 file path. In that directory, the public and private keypairs are generated for connecting to AWS, service logs, a local repository of packages for recipes and artifacts, and a directory for past and present deployments pushed to this device. Feel free to explore the directory at /greengrass/v2 to get familiar with what is stored on the device; though, you will need to escalate permissions using sudo to browse everything.

The installation added the first **component** to your IoT Greengrass environment, named `aws.greengrass.Nucleus`. The nucleus component is the foundation of IoT Greengrass; it is the only mandatory component, and it facilitates key functionality such as deployments, orchestration, and life cycle management for all other components. Without the nucleus component, there is no IoT Greengrass.

Additionally, the installation created the first **deployment** made to your device through the use of the `--deploy-dev-tools true` argument. That deployment installed a component named `aws.greengrass.Cli`. This second component includes a script, called `greengrass-cli`, that is used for local development tasks such as reviewing deployments, components, and logs. It can also be used to create new components and deployments. Remember, with IoT Greengrass, you can work locally on the device or push deployments remotely to it through AWS. Remote deployments are introduced in *Chapter 4, Extending the Cloud to the Edge*.

In AWS, a few different resources were created. First, a new **thing** was added to the **thing registry** in IoT Core. A thing is a logical representation of a physical device to which credentials, metadata, and other configuration are attached. The name of the created thing is `hbshub001` from the IoT Greengrass provisioning argument, `--thing-name`. Similarly, a new **thing group** was also created in the registry, named `hbshubprototypes`, from the `--thing-group-name` provisioning argument. A thing group contains zero or more things and thing groups. The design of IoT Greengrass uses thing groups to identify sets of edge devices that should have the same deployments running on them. For example, if you provisioned another hub prototype device, you would add it to the same `hbshubprototypes` thing group such that the new prototype deployments would travel to all of your prototype devices.

Additionally, your `hbshub001` thing has an entity attached to it called a **certificate**. The certificate is the record stored in IoT Core that was generated with the **x.509 public and private keypair** by the installation of IoT Greengrass. The keypair is stored on your device in the `/greengrass/v2` directory and is used to establish mutually authenticated connections to AWS. The certificate is how AWS recognizes the device when it connects with its private key (the certificate is attached to the `hbshub001` thing record) and knows how to look up permissions for the device. Those permissions are defined in another resource called the **IoT policy**.

An IoT policy is similar to an AWS IAM policy in that it defines explicit permissions for what an actor is allowed to do when interacting with AWS. In the case of an IoT policy, the actor is the device and permissions are for actions such as opening a connection, publishing and receiving messages, and accessing static resources defined in deployments. Devices get their permissions through their certificate, meaning a thing is attached to the certificate, and the certificate is attached to one or more policies. Here's a sketch of how these basic resources are related in the edge and the cloud:

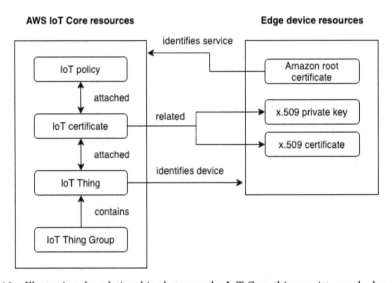

Figure 2.10 – Illustrating the relationships between the IoT Core thing registry and edge resources

In the cloud service of IoT Greengrass, a few more resources were defined for the initial provisioning of your device and its first deployment. An IoT Greengrass **core** is a mapping of a device (which is also known as a thing), the components and deployments running on the device, and the associated thing group the device is in. Additionally, a core stores metadata such as the version of IoT Greengrass core software installed and the status of the last known health check. Here is an alternate view of the relationship graph with IoT Greengrass resources included:

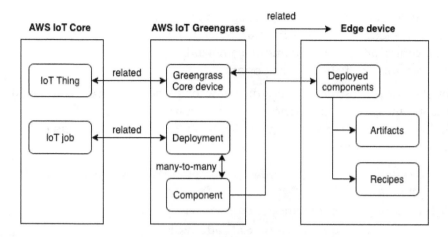

Figure 2.11 – Illustrating the relationships between IoT Core, IoT Greengrass, and the edge device

Now that you have installed IoT Greengrass and have an understanding of the resources created on provisioning, let's review what a component looks like when deployed to inform your implementation of the *Hello, world* component.

Creating your first edge component

The most basic milestone of any developer education is the *Hello, world* example. For your first edge component deployed to IoT Greengrass, you will create a simple *Hello, world* application in order to reinforce concepts such as component definition, a dependency graph, and how to create a new deployment.

Reviewing an existing component

Before you get started with drafting a new component, take a moment to familiarize yourself with the existing components that have already been deployed by using the IoT Greengrass CLI. This CLI was installed by the `--deploy-dev-tools true` argument that was passed in during the installation. This tool is designed to help you with a local development loop; however, as a best practice, it is not installed in production solutions. It is installed at `/greengrass/v2/bin/greengrass-cli`. The following steps demonstrate how to use this tool:

1. Try invoking the `help` command. In the Terminal app of your edge device, run `/greengrass/v2/bin/greengrass-cli help`.

2. You should view the output of the `help` command, including references to the `component`, `deployment`, and `logs` commands. Try invoking the `help` command on the `component` command: `/greengrass/v2/bin/greengrass-cli help component`.

3. You should view instructions regarding how to use the `component` command. Next, try invoking the `component list` command to show all of the locally installed components, `/greengrass/v2/bin/greengrass-cli component list`:

```
pi@raspberrypi:~ $ /greengrass/v2/bin/greengrass-cli
component list
java.lang.RuntimeException: Unable to create ipc client
    at com.aws.greengrass.cli.adapter.
impl.NucleusAdapterIpcClientImpl.
getIpcClient(NucleusAdapterIpcClientImpl.java:260)
...
    at com.aws.greengrass.cli.CLI.main(CLI.java:57)
Caused by: java.io.IOException: Not able to find auth
information in directory: /greengrass/v2/cli_ipc_info.
Please run CLI as authorized user or group.
```

Uh oh, what happened in this step? This command failed with a Java stack trace and errors such as `java.lang.RuntimeException: Unable to create ipc client` and `Please run CLI as authorized user or group`. This is an example of IoT Greengrass security principles at work. By default, the IoT Greengrass software is installed as the root system user. Only the root user, or a system user added to `/etc/sudoers`, can interact with the IoT Greengrass core software, even through the IoT Greengrass CLI. Components are run as the default system user identified in the configuration (please refer to the `--component-default-user` argument in the installation command), or each component can define an override system user to run as. Run the command again using `sudo` (*superuser do*):

4. `sudo /greengrass/v2/bin/greengrass-cli component list`

> **Note**
>
> There is a shortcut you can add to avoid typing the full path of /
> greengrass/v2/bin/greengrass-cli every time you want to
> use the CLI. You can add the /greengrass/v2/bin directory to your
> sudoers secure_path so that the greengrass-cli script can be
> used by typing in the name of the script. Use visudo to append the path to
> the Defaults secure_path list. This results in the use of a CLI such as
> sudo greengrass-cli component list as your system user in
> the sudo group.

Now you can view the list of components, including the aws.greengrass.
Cli component that installed the very CLI you are using! The CLI script can
be run by any system user; however, it will only successfully interact with the
local IoT Greengrass installation when run as root (via sudo) or a system
user belonging to a system group defined in the component's configuration
AuthorizedPosixGroups (which defaults to null). Note that even the nucleus
component appears in the list:

```
Component Name: aws.greengrass.Cli
    Version: 2.4.0
    State: RUNNING
    Configuration: {"AuthorizedPosixGroups":null}
```

The preceding output shows you the status of the component as a brief summary.
You can view the current state the component is in; in this case, this is RUNNING,
indicating the component's life cycle is either on or available. That makes sense
because the CLI should always be available to us while the component is being
deployed. Components that run once, perform a task, and close would show a state
of FINISHED after they have completed their life cycle tasks.

That's just the component's status, so next, let's take a look at the component's recipe
and artifacts. As defined earlier, a component is made up of two resources: a recipe
file and a set of artifacts. So what does the recipe file for the CLI component look
like? You can find this in the /greengrass/v2/packages/recipes directory.

5. You don't need to run the following *A* and *B* commands. They are included here to show you how to find the file contents later on:

A. To find the recipe file, use `sudo ls /greengrass/v2/packages/recipes/`.

B. To inspect the file, use `sudo less /greengrass/v2/recipes/rQVjcR-rX_XGFHg0WYKAnptIez3HKwtctL_2BKKZegM@2.4.0.recipe.yaml` (note that your filename will be different):

```
Selection of recipe.yaml for aws.greengrass.C--
RecipeFormatVersion: "2020-01-25"
ComponentName: "aws.greengrass.Cli"
ComponentVersion: "2.4.0"
ComponentType: "aws.greengrass.plugin"
ComponentDescription: "The Greengrass CLI component
provides a local command-line interface that you
can use on Greengrass core devices to develop and
debug components locally. The Greengrass CLI lets you
create local deployments and restart components on the
Greengrass core device, for example."
ComponentPublisher: "AWS"
ComponentDependencies:
  aws.greengrass.Nucleus:
    VersionRequirement: ">=2.4.0 <2.5.0"
    DependencyType: "SOFT"
Manifests:
- Platform:
    os: "linux"
  Lifecycle: {}
  Artifacts:
  - Uri:
"greengrass:UbhqXXSJj65QLVH5UqL6nBRterSKIhQu5FKeVAStZGc=/
aws.greengrass.cli.client.zip"
    Digest: "uziZS73Z6dKgQgB0tna9WCJ1KhtyhsAb/
DSv2Eaev8I="
    Algorithm: "SHA-256"
    Unarchive: "ZIP"
    Permission:
      Read: "ALL"
```

```
        Execute: "ALL"
    - Uri: "greengrass:2U_cb2X7-GFaXPMsXRutuT_
  zB6CdImClH0DSNVvzy1Y=/aws.greengrass.Cli.jar"
      Digest: "UpynbTgG+wYShtkcAr3X+l8/9QerGwaMw5U4IiicrMc="
      Algorithm: "SHA-256"
      Unarchive: "NONE"
      Permission:
        Read: "OWNER"
        Execute: "NONE"
  Lifecycle: {}
```

There are a few important observations to review in this file:

- Component names use a reverse domain scheme that is similar to the namespacing Java package. Your custom components in this book's project will start with com.hbs.hub, signifying components written for the Home Base Solutions hub product.

- This component is tied to specific versions of the IoT Greengrass nucleus, which is why the version is 2.4.0. Your components can specify any version here, and the best practice is to follow the semantic versioning specification.

- The ComponentType property is only used by AWS plugins such as this CLI. Your custom components will not define this property.

- This component only works with a specific version of the nucleus, so it defines a soft dependency on the aws.greengrass.nucleus component. Your custom components do not need to specify a nucleus dependency by default. This is where you will define dependencies on other components, for example, a component that ensures Python3 is installed before loading a component with a Python application.

- This component defines no specific life cycle activities, either at the global level or specific to the linux platform version of the manifest.

- The artifacts defined are for specific IoT Greengrass service files. You can view these files on disk in the /greengrass/v2/packages/artifacts directory. Your artifact URIs will use the s3://path/to/my/file pattern when deploying them from the cloud. During local development, your manifest does not need to define artifacts, as they are expected to already be on disk.

- Note the permissions on the two artifacts. The ZIP file can be read by any system user. In comparison, the JAR file can only be read by the OWNER, which in this scenario, means the default system user that was defined at installation, for example, the ggc_user user.

With this review of the component structure, it's time to write your own component with an artifact and a recipe.

Writing your first component

As mentioned earlier, the first component that we wish to create is a simple "Hello, world" application. In this component, you will create a shell script that prints `Hello, world` with the `echo` command. That shell script is an artifact for your component. Additionally, you will write a recipe file that tells IoT Greengrass how to use that shell script as a component. Finally, you will use the local IoT Greengrass CLI to deploy this component and check it works.

Local component development uses artifacts and recipe files available on the local disk, so you will need to create some folders for your working files. There is no folder in `/greengrass/v2` that is designed to store your working files. Therefore, you will create a simple folder tree and add your component files there:

1. From the Terminal app of your edge device, change the directory to your user's home directory: `cd ~/`.

2. Create a new folder to hold your local component resources: `mkdir -p hbshub/ {artifacts,recipes}`.

3. Next, create the path for a new artifact and add a shell script to its folder. Let's choose the component name of `com.hbs.hub.HelloWorld` and start the version at 1.0.0. Change the directory to the artifacts folder: `cd hbshub/ artifacts`.

4. Make a new directory for your component's artifacts: `mkdir -p com.hbs.hub. HelloWorld/1.0.0`.

5. Create a new file for the shell script: `touch com.hbs.hub. HelloWorld/1.0.0/hello.sh`.

6. Give this file write permissions: `chmod +x com.hbs.hub. HelloWorld/1.0.0/hello.sh`.

7. Open the file in an editor: `nano com.hbs.hub.HelloWorld/1.0.0/ hello.sh`.

8. Inside this editor, add the following content (this is also available in this chapter's GitHub repository):

hello.sh

```bash
#!/bin/bash

if [ -z $1 ]; then
        target="world"
else
        target=$1
fi

echo "Hello, $target"
```

9. Test your script with and without an argument. The script will print `Hello, world` unless provided with an argument to substitute for `world`:

    A. `./com.hbs.hub.HelloWorld/1.0.0/hello.sh`

    B. `./com.hbs.hub.HelloWorld/1.0.0/hello.sh friend`

10. That's all you need for your component's artifact. Next, you will learn how to take advantage of passing in arguments from inside the recipe file. Change the directory to the recipes directory: `cd ~/hbshub/recipes`.

11. Open the editor to create the recipe file: `nano com.hbs.hub.HelloWorld-1.0.0.json`.

12. Add the following content to the file. You can also copy this file from the book's GitHub repository:

com.hbs.hub.HelloWorld-1.0.0.json

```json
{
    "RecipeFormatVersion": "2020-01-25",
    "ComponentName": "com.hbs.hub.HelloWorld",
    "ComponentVersion": "1.0.0",
    "ComponentDescription": "My first AWS IoT Greengrass component.",
    "ComponentPublisher": "Home Base Solutions",
    "ComponentConfiguration": {
```

```
      "DefaultConfiguration": {
        "Message": "world!"
      }
    },
    "Manifests": [
      {
        "Platform": {
          "os": "linux"
        },
        "Lifecycle": {
          "Run": ". {artifacts:path}/hello.sh
'{configuration:/Message}'"
        }
      }
    ]
  }
```

This recipe is straightforward: it defines a life cycle step to run our `hello.sh` script that it will find in the deployed artifacts path. One new addition that has not yet been covered is the component configuration. The `ComponentConfiguration` object allows developers to define arbitrary key-value pairs that can be referenced in the rest of the recipe file. In this scenario, we define a default value to pass as an argument to the script. This value can be overridden when deploying a component to customize how each edge device uses the deployed component.

So, how do you test a component now that you've written the recipe and provided the artifacts? The next step is to create a new deployment that tells the local IoT Greengrass environment to load your new component and start evaluating life cycle events for it. This is where the IoT Greengrass CLI can help.

13. Use the following command to create a new deployment that includes your new component:

```
sudo /greengrass/v2/bin/greengrass-cli deployment
create    --recipeDir ~/hbshub/recipes --artifactDir ~/
hbshub/artifacts --merge "com.hbs.hub.HelloWorld=1.0.0"
```

14. You should view a response similar to the following:

```
Local deployment submitted! Deployment Id: b0152914-869c-
4fec-b24a-37baf50f3f69
```

15. You can verify that the component was successfully deployed (and has already finished running) with `sudo /greengrass/v2/bin/greengrass-cli component list`:

```
Components currently running in Greengrass:
Component Name: com.hbs.hub.HelloWorld
    Version: 1.0.0
    State: FINISHED
    Configuration: {"Message":"world!"}
```

16. You can view the output of this component in its log file: `sudo less /greengrass/v2/logs/com.hbs.hub.HelloWorld.log` (remember, the `/greengrass/v2 directory` is owned by root, so the log files must also be accessed with `sudo`):

```
2021-05-26T22:22:02.325Z [INFO] (pool-2-
thread-32) com.hbs.hub.HelloWorld: shell-runner-
start. {scriptName=services.com.hbs.hub.HelloWorld.
lifecycle.Run, serviceName=com.hbs.hub.HelloWorld,
currentState=STARTING, command=["/greengrass/v2/
packages/artifacts/com.hbs.hub.HelloWorld/1.0.0/hello.sh
'world..."]}
2021-05-26T22:22:02.357Z [INFO] (Copier) com.hbs.hub.
HelloWorld: stdout. Hello, world!. {scriptName=services.
com.hbs.hub.HelloWorld.lifecycle.Run, serviceName=com.
hbs.hub.HelloWorld, currentState=RUNNING}
2021-05-26T22:22:02.365Z [INFO] (Copier) com.hbs.
hub.HelloWorld: Run script exited. {exitCode=0,
serviceName=com.hbs.hub.HelloWorld, currentState=RUNNING}
```

Congratulations! You have written and deployed your first component to your Home Base Solutions prototype hub using IoT Greengrass. In the log output, you can observe two noteworthy observations. First, you can view the chronology of the component's life cycle change state from STARTING to RUNNING before reporting a successful exit code back to IoT Greengrass. The component concludes at that point, so we don't view an entry in the log that shows it move to the FINISHED state, although that is visible in the `greengrass.log` file.

Second, you can view the message written to STDOUT with an exclamation point included (world!). This means that the script received your component's default configuration instead of falling back on the default built into hello.sh (world). You could also override the default configuration value of "world!" in the recipe file with a custom value included in the deployment command. You'll learn how to use that technique to configure fleets in *Chapter 4, Extending the Cloud to the Edge.*

Summary

In this chapter, you learned the basics regarding a specific tool we will use throughout this book that satisfies one of the key needs of any edge ML solution, that is, the runtime orchestrator. IoT Greengrass provides out-of-the-box features to focus developers on their business solutions instead of the undifferentiated work to architect a flexible, resilient edge runtime and deployment mechanism. You learned that the fundamental unit of software in IoT Greengrass is the component, which is made up of a recipe and a set of artifacts, and components make their way to the solution via deployments. You learned how to validate that a device is ready to work with IoT Greengrass using the IDT. You learned how to install IoT Greengrass, develop your first component, and get it running in the local environment.

In the next chapter, we will take a deeper dive into how IoT Greengrass works by exploring how it enables gateway functionality, common protocols used at the edge, security best practices, and builds out new components used to sense and actuate in a cyber-physical solution.

Knowledge check

Before moving on to the next chapter, test your knowledge by answering these questions. The answers can be found at the end of the book:

1. Which of the following is the best practice for how to organize code in edge ML solutions? A monolithic application or isolated services?

2. What is the benefit of decoupling services in your edge architecture?

3. What is the benefit of isolating your code and dependencies from other services?

4. What is one trade-off to consider when choosing between wired and wireless networking implementations in IoT solutions?

5. What is an example of a smart home device that uses both a sensor and an actuator?

6. What are the two kinds of resources that define an IoT Greengrass component?

7. True or false: A component must define at least one artifact in its recipe.

8. Why is it a good design principle that, by default, only the root system user can interact with files in the IoT Greengrass directory?

9. True or false: Components can be deployed to IoT Greengrass devices either locally or remotely.

10. Can you think of three different methods that you could use to update the behavior of your `Hello, world` component to print `Hello, Home Base Solutions customer!`?

References

Please refer to the following resources for additional information on the concepts discussed in this chapter:

- The semantic versioning specification at `https://semver.org`.

- *Service-Oriented Architecture: Analysis and Design for Services and Microservices* by Erl Thomas, Pearson, 2016.

- *Foundations of Analog and Digital Electric Circuits* by Anant Agarwal, Jeffrey H. Lang, and Morgan Kaufmann, 2005.

3
Building the Edge

In this chapter, you will learn about **edge** solution concepts such as gateways and how **AWS IoT Greengrass** is used as a powerful edge appliance to interact with physical interfaces and leaf devices. The goal of this chapter is to start building proficiency with the use of IoT Greengrass for the writing and deploying of software components. This material is foundational to much of the book's hands-on projects and for understanding how we build solutions for the edge.

We will introduce you to the different protocols that IoT Greengrass can support out of the box and discuss commonly used protocols when building edge solutions. Additionally, we will review several security best practices for you to learn how to keep your edge workloads protected from threats and vulnerabilities. The chapter concludes with a hands-on activity to connect your first two device capabilities as components, whether using actual hardware or a pair of simulators.

In this chapter, we're going to cover the following main topics:

- Exploring the topology of the edge

- Reviewing common standards and protocols

- Security at the edge

- Connecting your first device – sensing at the edge

- Connecting your second device – actuating at the edge

Technical requirements

To complete the hands-on exercises in this chapter, you will need to have completed the steps in *Chapter 2, Foundations of Edge Workloads* such that your edge device has been set up with the IoT Greengrass Core software running and the `greengrass-cli` component installed.

You will want to clone the chapter's resources from the book's GitHub repository, for ease of use, if you haven't already done so. There is a step included in the *Connecting your first device – sensing at the edge* section that enables you to clone the repository at `https://github.com/PacktPublishing/Intelligent-Workloads-at-the-Edge/tree/main/chapter3`. You can perform this step now if you would like to browse the resources in advance:

```
git clone https://github.com/PacktPublishing/Intelligent-
Workloads-at-the-Edge-
```

As a reminder, the hands-on steps for this book were authored with a **Raspberry Pi** and **Sense HAT** expansion board in mind. For those of you using other Linux-based systems for the edge device, alternate technical resources are included in the GitHub repository with guidance on how to substitute them.

Exploring the topology of the edge

Solutions built for the edge take on many shapes and sizes. The number of distinct devices included in a solution ranges from one to many. The network layout, compute resources, and budget allowed will drive your architectural and implementation decisions. In an edge **machine learning** (**ML**) solution, we should consider the requirements for running ML models. ML models work more accurately when they are custom built for a specific instance of a device, as opposed to one model supporting many physical instances of the same device. This means that as the number of devices supported by an edge ML workload grows, so too will the number of ML models and compute resources required at the edge. There are four topologies to consider when architecting an edge ML solution: star, bus, tree, and hybrid. Here is a description of each of them:

- **Star topology**: The **Home Base Solutions** (**HBS**) hub device and appliance monitoring kits represent a common pattern in edge ML solutions called **star topology**. The appliance monitoring kits are single-purpose devices that report telemetry back to the hub device. This creates several advantages in terms of cost optimization for the kits since they do not need to bundle all of the hardware that is necessary to independently communicate directly with a cloud solution. Nor do they require any compute power or local storage to run their own ML models. The hub device in the star topology acts as a server in the sense that it can exchange data with the kits and perform heavier compute and storage tasks on their behalf. Entities such as leaf devices or software components address other entities directly to send them messages, and they get routed to the right destination. The following diagram shows an example of the HBS product design operating in a star topology:

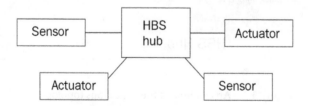

Figure 3.1 – An example of the star topology with hub and devices

- **Bus topology**: In a **bus** pattern, all of the nodes in the topology are connected to a single shared line, meaning a message published by one entity can be received by as many entities as are sharing the bus. The bus pattern comes from computer network history where devices are used to physically tap into the network line, expanding the bus with each device. While these days, we don't usually view this pattern as physically wired, there is a logical application of the bus pattern in edge ML solutions. In a decoupled solution, such as the one we are building, an entity such as a software component or leaf device can publish a message without addressing any other particular entity by using a topic address. A topic address doesn't strictly identify other entities; it is up to those other entities to subscribe to such topic addresses in order to get a copy of the message. In this way, the hub device is, technically, the center of a star-like topology; however, the way in which connected entities interact with each other is, in practice, more like a bus. The following diagram illustrates the bus concept for a parallel universe where HBS delivers the monitoring of industrial equipment with an equipment monitoring kit, a local server running ML inference, and an **andon** light all connected to a hub:

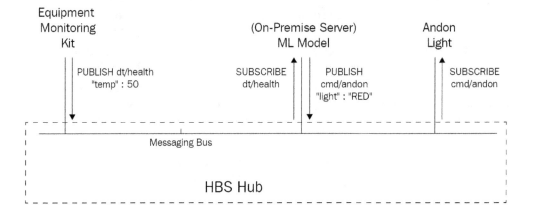

Figure 3.2 – An example of the logical bus topology

- **Tree topology**: While it is a good practice for any edge ML solution to functionally operate in isolation, we cannot ignore the benefits of bridging our solution to the wider network and cloud services. In a **tree** topology, our hub device is just one layer of nodes in a tree graph where a centralized service communicates with our fleet of hubs. Each hub is then responsible for a specific number of leaf devices and components running in a local star pattern. Managing our HBS product at scale requires us to think about the fleet in its entirety. The following diagram shows the relationship between a cloud service orchestrating our fleet, the fleet of HBS hub devices, and the local appliance monitoring kits supported per hub:

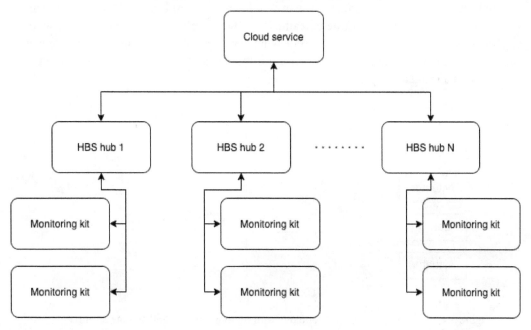

Figure 3.3 – An example of the tree topology for managing a fleet of connected hubs

- **Hybrid topology**: If our product design or hub device budget didn't allow running ML workloads at the edge and simply handled the cloud connectivity on behalf of the kits, this would necessitate a **hybrid** topology. In a hybrid topology, the hub might centralize just enough resources to establish the cloud connectivity and routes messages back and forth between the kits and the cloud service. Here, hybrid defines the additional topological complexity of running further compute workloads, such as our ML inference, in the cloud. In this scenario, the cloud-based ML workloads making inferences against incoming telemetry would require some subset of device messages to be transmitted to the cloud. Some scenarios might opt to reduce the bill of materials of the hub in favor of a cloud-based ML solution. This makes sense when the volume of traffic is on the lower end of the spectrum or when the number of ML workloads exceeds what is reasonable to run on a single gateway device. The following diagram shows a modified example of our fictional product design running as a hybrid:

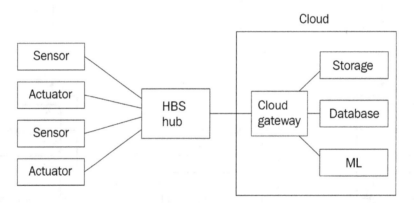

Figure 3.4 – An example of hybrid topology with remote compute and storage resources

There are two additional patterns that are common when studying network topologies, that is, the mesh and ring topologies:

- In a **mesh topology**, nodes can have one-to-many relationships with other nodes and exchange information through that network of connections to reach that information's destination:

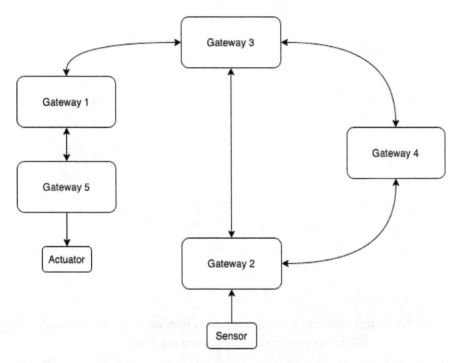

Figure 3.5 – A mesh topology where a sensor on gateway 2 traverses the mesh to reach an actuator on gateway 5

- In a **ring topology**, nodes have at most two neighboring connections and exchange information through the ring until it reaches its destination:

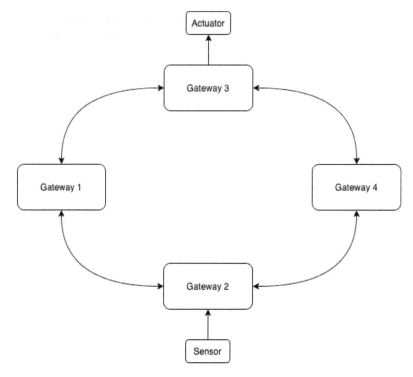

Figure 3.6 – A ring topology where a sensor reading from gateway 2 passes through adjacent gateways to reach an actuator at gateway 3

These patterns emphasize decentralization where nodes connect to each other directly. While these patterns have their time and place in the broader spectrum of IoT solutions, they are infrequently used in edge ML solutions where a gateway or hub device and cloud service are often best practices or outright requirements.

When deciding on a particular topology for your solution architecture, start by considering whether all devices at the edge are weighted equally or whether they will communicate with a central node such as a gateway. A consumer product design for an edge ML solution tends to use the star pattern when thinking about the edge in isolation. A good edge solution should be able to operate in its star pattern even when any external link to a larger tree or hybrid topology is severed. We use the star pattern to implement the HBS product since the hub device will run any and all ML runtime workloads that we require.

IoT Greengrass is designed to run as the hub of a star implementation and plug into a larger tree or hybrid topology connecting to the AWS cloud. Solution architects can choose how much or how little compute work is performed by the gateway device running IoT Greengrass. In the next section, we will review the protocols used to exchange messages at the edge and between the edge and cloud.

Reviewing common standards and protocols

Protocols define the specifications for exchanging messages with an edge solution. This means the format of the message, how it is serialized over the wire, and also the networking protocols for communicating between two actors in the solution. Standards and protocols help us to architect within best practices and enable quick decision-making when selecting implementations. Before diving into the common protocols that are used in edge solutions, first, let's review two architectural standards used in information technology and operations technology to gain an understanding of where IoT Greengrass fits into these standards. Using these as a baseline will help set the context for the protocols used and how messages traverse these models in an edge solution.

IoT Greengrass in the OSI model

The **Open Systems Interconnection (OSI)** model defines a stack of seven layers of network communications, describing the purpose and media used to exchange information between devices at each layer. At the top of the stack is layer seven, the *application layer*, where high-level APIs and transfer protocols are defined. At the bottom is layer one, the *physical layer*, where digital bits are transmitted over physical media using electricity and radio signals. The following is a diagram of the OSI model and shows how IoT Greengrass fits in with individual layers:

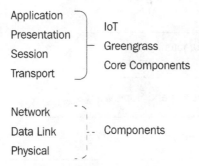

Figure 3.7 – An OSI model with an overlay of IoT Greengrass layer interactions

Here, you can observe that our runtime orchestrator, IoT Greengrass, operates from layer four to layer seven. There are high-level applications and transfer protocols used in the IoT Greengrass Core software to exchange application messages with the AWS cloud and local devices using protocols such as HTTPS and MQTT. Additionally, libraries bundled in the core software are responsible for the transport layer communications in the TCP/IP stack, but then further transmission throughout the OSI model is handed off to the host operating system.

Note that while the IoT Greengrass Core software operates from layer four to layer seven, the software components deployed to your edge solution might reach all the way down to layer one. For example, any sensors or actuators physically connected to the IoT Greengrass device could be accessed with code running in a component, usually with a low-level library API.

IoT Greengrass in ANSI/ISA-95

American National Standards Institute/International Society of Automation standard 95 (ANSI/ISA-95) defines the process in which to relate the interfaces between the enterprise and control systems. This standard is used in enterprise and industrial solution architectures. It describes another layered hierarchy; this one is for the level at which systems are controlled and suggests the time scale at which decisions must be made. The following diagram presents another frame of reference for how IoT Greengrass and an edge ML solution fit into a holistic picture:

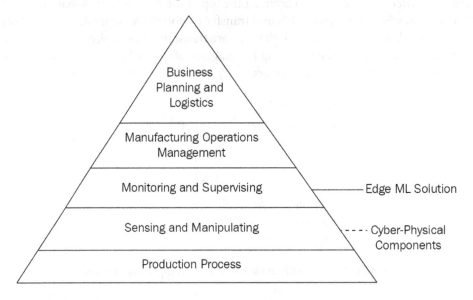

Figure 3.8 – The ISA-95 model with an overlay of IoT Greengrass layer interactions

Here, you can observe that IoT Greengrass primarily fits in layer three, the *Monitoring and Supervising* layer of control systems, to facilitate the upward aggregation of device telemetry, downward distribution of commands, and handle some decision making in a supervisory capacity. IoT Greengrass is useful in any kind of edge solution, be it consumer-grade products or to facilitate the operation of industrial machinery. While our HBS product example is not an industrial use case, the same pattern applies in that our hub device performs as a gateway for sensor monitoring equipment.

Now that you have a framework regarding how IoT Greengrass fits into these hierarchies, we can review common protocols that are used at the relevant layers.

Application layer protocols

Examples of application layer communications include requesting data from an API, publishing sensor telemetry, or sending a command to a device. This kind of data is relevant to the solution itself and the business logic in service of your solution's outcomes. For example, none of the other layers of the OSI model, such as the transport layer or the network layer, make decisions in the event that a deployed sensor is reporting the ambient temperature at 22°C. Only the running applications of your solution can use this data and send or receive that data by interacting with the application layer.

To communicate between the edge and the cloud, the most popular application layer protocol is **HTTPS** for request-response interactions. IoT Greengrass uses HTTPS to interact with AWS cloud services for the purposes of fetching metadata and downloading resources for your components, such as the component recipe and artifacts such as your code and trained ML models. Additionally, your custom components running at the edge might use HTTPS to interact with other AWS services, on-premises systems, and the APIs of other remote servers.

To exchange messages between the edge and the cloud, and within the edge solution, bi-directional messaging protocols (also called *publish-subscribe* or *pubsub*) are commonly used, such as MQTT or AMQP. The benefits of these kinds of protocols are listed as follows:

- They use stateful connections to minimize the frequency of handshake connections.
- Traffic can flow in either direction without one end or the other having to be responsible for initiating a new exchange.
- They offer minimal overhead per message making them ideal for constrained devices.
- Clients at the edge can initiate these connections, eliminating the need for inbound connections to be permitted by network firewall rules.

IoT Greengrass uses the MQTT protocol to open connections to the AWS IoT Core service in a client-broker model in order to pass messages from local devices up to the cloud, receive commands from the cloud and relay them to local devices, and synchronize the state after a period of disconnection. Additionally, IoT Greengrass can serve as the broker to other local devices that connect to it via MQTT. The following is a diagram of an IoT Greengrass device, such as the HBS hub device, acting as both the client and the broker:

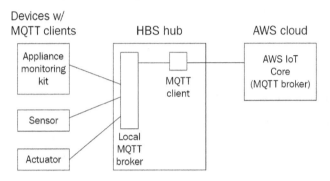

Figure 3.9 – The HBS hub acting as both a client to AWS IoT and a server to local devices

Next up are the message format protocols that describe the way data is structured as it is sent over the application layer protocols.

Message format protocols

If a messaging protocol such as MQTT specifies how connections are established and how messages are exchanged, a *message format protocol* specifies what the shape and content of an exchanged message are. You can consider a telephone as an analogy. The telephone handset represents how speech is sent in both directions, but the language being spoken by the participants at both ends must be understood in order for that speech to make sense! In this analogy, MQTT represents the telephone itself (abstracting away the public telephone exchange network), and the message format protocol is the language being spoken by the people on either end.

For any two participants exchanging data with each other, that data is either transmitted as binary or it will go through a process of serialization and deserialization. Common message format protocols used in edge solutions include **JavaScript Object Notation (JSON)**, Google **Protocol Buffers (protobuf)**, and **Binary JSON (BSON)**. These formats make it easier for devices, edge components, and cloud solutions to interoperate. This is especially important in an architecture that is inclusive of mixed programming languages. The message format is a means of abstraction that is key to architecting solutions. By using a serializable message format protocol, the following diagram shows how a component written in Python can exchange messages with a component written in Java:

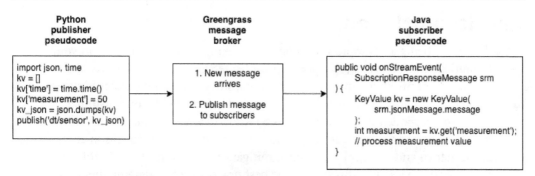

| Python
publisher
pseudocode | Greengrass
message
broker | Java
subscriber
pseudocode |

```
import json, time
kv = []
kv['time'] = time.time()
kv['measurement'] = 50
kv_json = json.dumps(kv)
publish('dt/sensor', kv_json)
```

```
1. New message
   arrives

2. Publish message
   to subscribers
```

```
public void onStreamEvent(
    SubscriptionResponseMessage srm
){
    KeyValue kv = new KeyValue(
        srm.jsonMessage.message
    );
    int measurement = kv.get('measurement');
    // process measurement value
}
```

Figure 3.10 – An example of serialization/deserialization using Greengrass components

You could send all messages as binary data, but you would end up with an overhead in each recipient that would need to figure out what to do with that data or enact strict conventions for what can be sent. For example, a sensor device that only ever sends a numerical measurement in degrees centigrade could just send the value as binary data. If that system never changes, there's limited value to adding notation and serializing it. The recipient on the other end can be hardcoded to know what to do with it, thus saving overhead on metadata, structure, and bandwidth. This works for rigid, static systems and for cases where transmission costs must be the top priority for optimization.

Unstructured data such as images, video, and audio is commonly sent as binary payloads but with an accompanying header indicating what kind of data it is. In an HTTP request, the *Content-Type* header will include a value such as *text/HTML* or a MIME type such as *video/MP4*. This header tells the recipient how to process the binary payload of that message.

The **interprocess communication (IPC)** functionality that is provided by IoT Greengrass to components to enable the exchange of messages between them supports the JSON message format along with the raw binary format. In this chapter, you will build two components that use IPC to pass JSON messages from one component to the other.

> **Note**
> Since IoT Greengrass does not prescribe any particular protocol to interact with edge devices and systems, you can easily implement components that include libraries to interact with any device and any protocol.

The key takeaway regarding protocols is that we can use common protocols for the same, or similar, advantages as we use a good architecture pattern. They are battle-tested, well-documented, easy to implement, and prevent us from getting lost in the cycles of implementation details where our time would be better spent on delivering outcomes. In the next section, we will cover, at a high level, the security threats for an edge ML solution and some best practices and tools for mitigating them.

Security at the edge

With IoT security being a hot topic and frequently making headlines, security in your edge ML solutions must be your top priority. Your leadership at HBS will never want to see their company or product name in the news for a story concerning a hacked device. Ultimately, security is about establishing and maintaining trust with your customer. You can use a threat modeling exercise such as STRIDE to analyze atomic actors in your edge system such as end devices, gateways, and software components to reason about worst-case scenarios and the minimum viable mitigation to prevent them. In this section, we will cover common security threats and the best practices for mitigating them.

End devices to your gateway

Let's start with the terminal segment in our edge ML solution along with the appliance monitoring kit itself and its connection to the hub device. The worst-case scenario for this segment is that an unhealthy appliance is mistakenly reported as healthy. If a customer installs the product and it fails to do the one thing it advertises, this will lose all customer trust in the product. To mitigate this scenario, the sensor readings from the monitoring kit must be authentic. This means we must prevent the hub device from receiving false measurements from a spoofed kit.

Here, the best practice is to use some form of secret material that only the kit and the hub device understand. A secret can be a pre-shared key in a symmetrical cryptographic model, or it could be a public key and private key pair in an asymmetrical cryptographic model. If the kit can sign measurements sent to the hub device with a secret, then only the hub device can read them, and it will understand that it could only come from the device that it's paired with. Similarly, the monitoring kit would only act on messages, such as a request to update firmware, if those messages were signed by a secret it recognizes from the paired hub device.

A safe design pattern for our pairing process between the monitoring kit and hub device is to task the customer with a manual step, such as a physical button press. This is similar to the Wi-Fi pairing method called **Wi-Fi Protected Setup** (**WPS**). If the customer has to manually start the pairing process, this means it is harder for an attacker to initiate communication with either the kit or the hub. If an attacker has physical access to the customer's premises to initiate pairing with their own device, this would be a much larger security breach that compromises our future product.

IoT Greengrass provides a component called *secret manager* to help with this use case. The secret manager component can securely retrieve secret materials from the cloud through the AWS Secrets Manager service. You can build workflows into your edge solution, such as the monitoring kit pairing process, to establish a cryptographically verifiable relationship between your devices.

The gateway device

The following list of risks and mitigations focus on the gateway device itself, which runs the IoT Greengrass Core software:

- **Secure connectivity to the cloud**: So, how do you ensure that the connection between the core device and the cloud service, such as AWS IoT Core, is secure? Well, here, the worst-case scenarios are that the messages exchanged have been accessed, tampered with, or delivered to a spoofed endpoint. The best practice, and the one built into IoT Greengrass, is to use a **public key infrastructure** (**PKI**) and mutual validation. PKI implements asymmetrical cryptography by generating private keys and public certificates. A recipient uses the public certificate to validate that the messages from the sender are authentic. In a mutual validation model, both ends of a connection use this structure to prove the authenticity of the other during the handshake. Devices that include a **Trusted Platform Module** (**TPM**), which comes with a private key securely stored in it, can generate certificates for use in PKI without ever exposing the private key.

- **Inbound network connections**: Any device on a network could be susceptible to receiving incoming connection requests. The worst-case scenario is that an attacker gains remote access to a system and starts probing the system or executing code. The best practice for establishing connections is to refuse inbound-initiated connections and rely on outbound, client-initiated connections instead.

- **Boot tampering**: So, what happens if a device is physically modified between the point of manufacture and when a customer receives it? Well, malicious code could be loaded onto the device to be executed when it is received by the customer and added to the network. To prevent any tampering of the device, design your hardware platform with (or build on top of existing platforms that use) a TPM. A TPM can be used to validate that an encrypted disk has not been modified between boot sequences.

Edge components

Next, we will move on to the components that are running in your edge solution on the IoT Greengrass Core device:

- **Component integrity on disk**: So, what happens if an attacker can access the component artifacts after they are deployed to the core device? The worst-case scenario is that valuable intellectual property is stolen or modified to change component behavior. In IoT Greengrass, all component resources such as artifacts and recipes are stored on disk as root. The working directories that components use for file I/O belong to the default component system user in the IoT Greengrass configuration or the override user that has been specified during deployment. The best practice is to protect the root and component system user access by limiting additional system users beyond what is needed for the components of the solution. If additional system users are required to be on the device, they should not belong to the same groups as your component users or have permissions to escalate privilege to the root configuration.

- **Component integrity in the cloud**: So, what happens if an attacker modifies the component artifacts before they are retrieved by IoT Greengrass for a deployment? The worst-case scenario is that a tampered component uses the `RequiresPrivilege` flag and then has full access to the core device. IoT Greengrass prevents this attack by computing a digest (that is, a mathematical summation for asserting the authenticity of a payload) whenever you upload and register a component. The core device will validate the digest against the component artifacts and fail if an artifact has been tampered with.

- **Model poisoning**: So, what happens if an attacker compromises a model-backed decision-making process? For example, a camera feed using a local ML model to detect anomalous presence activity gets retrained with new data once a week. If an attacker understands the retraining loop, they can poison the training data over time with the intent of shifting the anomaly threshold to their advantage. To mitigate model poisoning, use **human-in-the-loop validation** to approve of new labeled data used in training. Human-in-the-loop validation means that you have a mechanism for your human experts to review unusual results, flagged results, or subsets of results as model quality assurance. Additionally, you can use a static approved training dataset to test later generations of models against their original success benchmark.

- **IPC**: When multiple components are deployed to an IoT Greengrass Core device, some of those components might need to exchange messages with each other over IPC. The worst-case scenario is that a subscribing component receives commands from a component that shouldn't be allowed to control it. Another scenario is that sensitive data is leaked to a component that isn't approved to have access to the data. IoT Greengrass provides a security control for the IPC functionality that allows you to specify explicit topics that components are allowed to use in publish and subscribe operations, respectively. A recommended practice is to design and document your topic tree, indicating which topics are sensitive. In this way, you can use automated review mechanisms to flag recipes in your development life cycle that use the * wildcard, or sensitive topics, for explicit approval.

So, this section covered a few high-risk security threats and the built-in mitigations provided by IoT Greengrass along with several best practices you can implement. Security at the edge is both complex and complicated. You can use threat modeling to identify the worst-case scenarios and best practices to mitigate those threats. In the next section, you will continue your journey as the HBS IoT architect by connecting two devices using components that deliver a simple sensor-to-actuator flow.

Connecting your first device – sensing at the edge

In this section, you will deploy a new component that delivers the first sensing capability of your edge solution. In the context of our HBS appliance monitoring kit and hub device, this first component will represent the sensor of an appliance monitoring kit. The sensor reports to the hub device the measured temperature and humidity of an attached **heating, ventilation, and air conditioning (HVAC)** appliance. Sensor data will be written to a local topic using the IPC feature of IoT Greengrass. A later section will deploy another component that consumes this sensor data.

If you are using a Raspberry Pi and a Sense HAT for your edge device, the temperature and humidity measurements will be taken from the Sense HAT board. For any other project configurations, you will use a software data producer component to simulate measurements of new data. Component definitions for both paths are available in the GitHub repository, in the `chapter3` folder.

Both versions of the component have been written for the Python 3 runtime and defined using Python virtual environments to isolate dependencies. You will deploy one or the other using `greengrass-cli` to create a new local deployment that merges with the component. This chapter covers steps regarding how to install the component that reads from and writes to the Sense HAT. Any procedural differences for the data producer and consumer components are covered in the GitHub repository's `README.md` file.

Installing the sensor component

Installing this component is just like installing the `com.hbs.hub.HelloWorld` component from *Chapter 2*, *Foundations of Edge Workloads*. You will use the IoT Greengrass CLI to merge in a predefined component using the `deployment` command:

1. On your hub device (Raspberry Pi), clone the Git repository that contains the book's companion resources:

    ```
    cd ~/ && git clone https://github.com/PacktPublishing/
    Intelligent-Workloads-at-the-Edge-.git
    ```

2. Change directory into the cloned folder for `chapter3`:

    ```
    cd Intelligent-Workloads-at-the-Edge-/chapter3
    ```

3. Create a deployment to include the `com.hbs.hub.ReadSenseHAT` component (or `com.hbs.hub.ReadSenseHATSimulated` if working on hardware other than a Raspberry Pi):

    ```
    sudo /greengrass/v2/bin/greengrass-cli deployment create
    --merge com.hbs.hub.ReadSenseHAT=1.0.0 --recipeDir
    recipes/ --artifactDir artifacts/
    ```

4. You can monitor the progress of the deployment in the log file:

    ```
    sudo tail -f /greengrass/v2/logs/greengrass.log
    ```

5. When the logs stop updating from the deployment, you can validate that the component was installed successfully:

    ```
    sudo /greengrass/v2/bin/greengrass-cli component list
    ```

6. You should observe the **RUNNING** status for `com.hbs.hub.ReadSenseHAT`.

Now that the component has been installed, let's review the component.

Reviewing the sensor component

Let's review some interesting bits from this sensor component so that you have a better idea of what's going on in this component. In this section, we will review a few highlights, starting with the recipe file.

IPC permissions

In the `com.hbs.hub.ReadSenseHAT-1.0.0.json` section, we are using a new concept in the configuration called `accessControl`. This configuration defines the features of IoT Greengrass that the component is allowed to use. In this case, the recipe is defining a permission to use IPC and publish messages to any local topic. The operation is `aws.greengrass#PublishToTopic`, and the resource is the * wildcard, meaning the component is permitted to publish to any topic. In a later section, you will observe a different permission defined here to subscribe to IPC and receive the messages being published by this component. Here is the relevant section of the recipe file showing the `accessControl` configuration:

com.hbs.hub.ReadSenseHAT-1.0.0.json

```
    "ComponentConfiguration": {
      "DefaultConfiguration": {
        "accessControl": {
          "aws.greengrass.ipc.pubsub": {
            "com.hbs.hub.ReadSenseHAT:pubsub:1": {
              "policyDescription": "Allows publish operations on
local IPC",
              "operations": [
                "aws.greengrass#PublishToTopic"
              ],
              "resources": [
                "*"
              ]
            }
          }
        }
      }
    },
```

In the preceding JSON snippet, you can see that the default configuration for this component includes the `accessControl` key. The first child of `accessControl` is a key that is used to describe which system permission we are setting. In this scenario, the permission is for the `aws.greengrass.ipc.pubsub` system. The next child key is a unique policy ID that must be unique across all of your components. The best practice is to use the format of *component name, system name or shorthand, and a counter*, all joined by colon characters. The list of operations includes just one permission for publishing messages, but it could also include the operation for subscribing. Finally, the list of resources indicates the explicit list of topics permitted for the preceding operations. In this scenario, we use the * wildcard for simplicity; however, a better practice for observing the principle of least privilege is to exhaustively enumerate topics.

Multiple life cycle steps

In the simple `"Hello, world"` component, there was just a single life cycle step to invoke the shell script at runtime. In this component, you can see that we are using two different life cycle steps: install and run. Each life cycle step is processed at a different event in the IoT Greengrass component life cycle:

- The **Bootstrap** step is evaluated when the component is first deployed or when a new version of the component is deployed. You should include instructions in the Bootstrap life cycle when you want Greengrass or the device to restart. This component doesn't require any restarts, so we exclude it from the recipe.

- The **Install** step will run each time the component is launched, for example, after any time the device restarts and Greengrass is restarting components. Here, you should include instructions that install or configure dependencies before your main component code starts.

- The **Run** step is only evaluated after the `Bootstrap` and `Install` scripts have been completed successfully. Use the `Run` step to run an application or one-off activity.

- Another kind of life cycle step is the **Startup** step and its matching **Shutdown** step. Use these steps to start and stop system background processes, services, and daemons. Note that you can only use either `Run` or `Startup` in a recipe, not both.

> **Note**
>
> The IoT Greengrass Core software supports multiple life cycle events in order to provide flexible use of the component recipe model and component dependency graph. A complete definition of these life cycle events can be found in the *References* section, which is at the end of the chapter. The Run, `Install`, and `Startup` life cycle events are the most commonly used in component recipes.

Let's take a closer look at the life cycle steps of this recipe:

com.hbs.hub.ReadSenseHAT-1.0.0.json

```json
"Lifecycle": {
        "Install": {
            "RequiresPrivilege": true,
            "Script": "usermod -a -G i2c,input ggc_user && apt
update && apt upgrade -y && apt install python3 libatlas-base-
dev -y && python3 -m venv env && env/bin/python -m pip install
-r {artifacts:path}/requirements.txt"
        },
        "Run": {
            "Script": "env/bin/python {artifacts:path}/read_
senseHAT.py"
        }
    }
```

In this recipe, we use the `Install` step to make system-level changes that require escalated permissions, such as making sure Python is installed. The `Install` step uses `pip` to install any Python libraries defined by the `requirements.txt` file in our component artifacts. Finally, the `Run` step invokes Python to start our program.

Virtual environments

In this Python component, we are using a feature of Python 3 called virtual environments. A virtual environment allows you to specify an explicit version of the Python runtime to use when interpreting code. We use this to install any dependency libraries locally, so neither the dependencies nor runtime conflict with the system-level Python. This reinforces the best practice of applying isolation to our component. The relative invocation of `env/bin/python` is telling the script to use the virtual environment's version of Python instead of the one at the system level at `/usr/bin/python`.

This component uses a `requirements.txt` file to store information about the Python packages used and the versions of those packages to install. The requirements file is stored as an artifact of the component, along with the Python code file. Since it is an artifact, the command to `pip` must use the `{artifacts:path}` variable provided by IoT Greengrass to locate this file on disk.

We could achieve even better isolation for our component in one of two ways:

- **System-level Python runtime management**: We could use a more specific methodology to load Python runtimes onto the device and reference the version this component needs. There is a risk of using the system-level Python 3 runtime that the recipe installs in the Bootstrap script since another component could later install a different Python 3 runtime. The best practice would be to use further components as dependencies to install each specific runtime that our component requires access to. In this way, a component such as this one could depend on a component that installs Python 3.7 and another component could depend on a component that installs Python 3.9, without conflicting with each other.

- **Containerization**: Containers are a piece of technology used to enforce even stricter process and dependency isolation than a Python virtual environment. We could build and deploy our sensor component in a Docker container that includes the Python runtime, system packages, and Python libraries and perform any additional custom steps in a container environment before invoking our Python code. This would achieve the best level of isolation; however, it has the drawback of increased complexity to develop and requires more total disk space consumption to achieve that level of isolation. For a production environment, you, as the IoT architect, are responsible for making trade-offs on how much isolation is warranted for the additional overhead.

Since this HBS project is a prototype and we are using a Raspberry Pi device that comes with Python 3 preinstalled, it is within acceptable risk to simply ensure Python 3 is installed. A comprehensive isolation approach with containers for every component could fit, but the lighter-weight approach with Python virtual environments is sufficient at this prototype stage.

Permissions to Unix devices

The code that reads from your Sense HAT device uses the Sense HAT Python library to read values from the device files that the Unix kernel exposes as device interfaces. These device files, such as /dev/i2c-1 and /dev/input/event2, are restricted to system users in groups such as i2c and input. The root user has access to these devices and a Raspberry Pi, and so does the default pi user. Our default component user, ggc_user, is not in these groups; therefore, code run as ggc_user will not be able to access values from these device interfaces. There are three ways to resolve this issue, which are listed as follows:

- First, you could add ggc_user to the i2c and input groups using a system command such as usermod -a -G i2c,input ggc_user.

- Second, you could set the `RequiresPrivilege` flag in the component recipe's Run script to `true`.

- Third, you could override which system user the component should run at deployment time by adding the `--runWith COMPONENT:posixUser=USERNAME` flag. (This can be done with a configuration change in the deployment component in production components, which is covered in *Chapter 4, Extending the Cloud to the Edge*.)

The best practice is to update the groups that the `ggc_user` component user is in. This reduces how often we use privileged access in our IoT Greengrass components and maintains our isolation principle by bundling the requirement in our recipe file. Running the component as the `pi` user isn't bad; however, as a developer, you should not assume this user will exist on every device and have the necessary permissions just because they are operating system defaults. For convenience, here is another clip of the life cycle step that sets up our user permissions for `ggc_user`:

com.hbs.hub.ReadSenseHAT-1.0.0.json

```
"Lifecycle": {
  "Install": {
    "RequiresPrivilege": true,
    "Script": "usermod -a -G i2c,input ggc_user && apt update
&& apt upgrade -y && apt install python3 libatlas-base-dev -y
&& python3 -m venv env && env/bin/python -m pip install -r
{artifacts:path}/requirements.txt"
  },
```

This covers the interesting new features used in the component recipe file. Next, let's take a deep dive into important bits of the code.

Logging

A critical part of monitoring your components is to log important events. These lines set up a logger object that you can use throughout your Python code. This should be standard in every application:

read_senseHAT.py

```
logger = logging.getLogger()
handler = logging.StreamHandler(sys.stdout)
```

```
logger.setLevel(logging.INFO)
logger.addHandler(handler)
```

When building Python applications for IoT Greengrass, you can copy lines such as these to Bootstrap logging. Note that your logger will capture logs at the INFO level or a level that is higher in criticality. Debug logs will not be captured unless you set the level to logging.DEBUG. You might set different levels of logs depending on where in the development life cycle you are, such as DEBUG in beta and INFO in production. You could set the logging level as a variable with component-level configuration and override it for a given deployment.

Reading from Sense HAT

Inside the build_message function is some simple code to initiate the Sense HAT interface and read values from its temperature and humidity sensors. These represent the values measured by our HBS appliance monitoring kit, attached to a fictional HVAC appliance:

Read_senseHAT.py

```
sense = SenseHat()
message = {}
message['timestamp'] = float("%.4f" % (time.time()))
message['device_id'] = 'hvac'
message['temperature'] = sense.get_temperature()
message['humidity'] = sense.get_humidity()
```

This code builds up a new object, called message, and sets child properties equal to the values we're getting from the Sense HAT library. The code also sets a simple device ID string, and generates the current timestamp.

Publishing a message

Next, we will cover the key lines of code inside the publish_message function:

read_senseHAT.py

```
publish_message = PublishMessage()
publish_message.json_message = JsonMessage()
publish_message.json_message.message = message
request = PublishToTopicRequest()
```

```
request.topic = topic
request.publish_message = publish_message
operation = ipc_client.new_publish_to_topic()
operation.activate(request)
future = operation.get_response()
try:
    future.result(TIMEOUT)
    logger.info('published message, payload is: %s', request.
publish_message)
except Exception as e:
    logger.error('failed message publish: ', e)
```

These lines of code prepare the message and the request that will be communicated to the IPC service of IoT Greengrass as a new publish operation. This code will look familiar in any later components that require you to publish messages to other components running on the HBS hub device.

Now that we have taken a tour of the sensor application source code, let's examine what values you are measuring in the log file.

Testing the sensor component

To inspect the values that you are sampling from the sensor, you can tail the log file for this component. If you are using the ReadSenseHATSimulated component, make sure you inspect that log file instead.

Tail the log file:

```
sudo tail -f /greengrass/v2/logs/com.hbs.hub.ReadSenseHAT.log
2021-06-29T01:03:07.746Z [INFO] (Copier) com.hbs.hub.
ReadSenseHAT: stdout. published message, payload is:
PublishMessage(json_message=JsonMessage(message={'timestamp':
1624928587.6789, 'device_id': 'hvac', 'temperature':
44.34784698486328, 'humidity': 22.96312713623047})).
{scriptName=services.com.hbs.hub.ReadSenseHAT.lifecycle.
Run.Script, serviceName=com.hbs.hub.ReadSenseHAT,
currentState=RUNNING}
```

You should observe new entries in the log file with the temperature and humidity measurements sampled. These values are being logged and also published over IPC to any other components that are listening for them. You don't have any other components listening on IPC yet, so now is a great time to move on to your second component.

Connecting your second device – actuating at the edge

The previously deployed component acts as a sensor to read values from a fictional appliance monitoring kit and publishes those values over IoT Greengrass IPC on a local topic. The next step is to create an actuator component that will respond to those published measurements and act upon them. Your actuator component will subscribe to the same local topic over IPC and render the sensor readings to the LED matrix of your Sense HAT board. For projects not using the Raspberry Pi with Sense HAT, the simulation actuator component will write measurements to a file as a proof of concept.

Installing the component

Similar to the previous installation, you will create a deployment that merges with the new component. Please refer to the earlier steps for the location of the source files and validation steps that the deployment concluded. For projects not using the Raspberry Pi with the Sense HAT module, you will deploy the com.hbs.hub.SimulatedActuator component instead.

Create a deployment to include the com.hbs.hub.WriteSenseHAT component:

```
sudo /greengrass/v2/bin/greengrass-cli deployment create
--merge com.hbs.hub.WriteSenseHAT=1.0.0 --recipeDir recipes/
--artifactDir artifacts/
```

Once deployed, you should start seeing messages appear on the LED matrix in the format of t: 40.15 h:23.79. These are the temperature and humidity values reported by your sensor component. The following photograph shows a snapshot of the LED matrix scrolling through the output message:

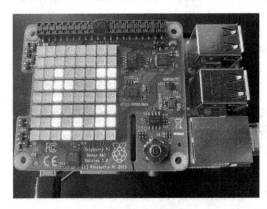

Figure 3.11 – A photograph of Sense HAT scrolling through reported sensor values

If, at any point, you get tired of seeing the scrolling messages on your device, you can simply remove the com.hbs.hub.WriteSenseHAT component with a new deployment, as follows:

```
sudo /greengrass/v2/bin/greengrass-cli deployment create
--remove com.hbs.hub.WriteSenseHAT
```

Read on to learn how this component is structured.

Reviewing the actuator component

Let's review the interesting differences between this component and the sensor component.

Recipe file differences

Starting with the recipe file, there is only one key difference to observe. Here is a snippet of the recipe file that we'll review:

com.hbs.hub.WriteSenseHAT-1.0.0.json

```
"accessControl": {
        "aws.greengrass.ipc.pubsub": {
          "com.hbs.hub.WriteSenseHAT:pubsub:1": {
            "policyDescription": "Allows subscribe operations
on local IPC",
            "operations": [
              "aws.greengrass#SubscribeToTopic"
            ],
            "resources": [
              "*"
            ]
          }
        }
    }
```

In the recipe for com.hbs.hub.WriteSenseHAT, the accessControl permission specifies a different operation, aws.greengrass#SubscribeToTopic. This is the inverse of what we defined in the sensor component (aws.greengrass#PublishToTopic). This permission allows the component to set up topic subscriptions on IPC and receive messages published by other IPC clients, such as the sensor component. The following diagram shows the contrast of IPC permissions between a publishing sensor and a subscribing actuator:

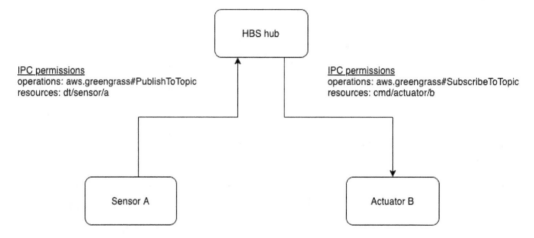

Figure 3.12 – IPC permissions for the sensor and the actuator

In addition to this, the resources list uses the * wildcard to indicate that the component can subscribe to any local topic. Following a principle of least privilege for a production solution, this list of resources would specify the explicit list of topics to which the component is allowed to publish or subscribe. Since this hub device is a prototype, the wildcard approach is acceptable. Each of the read and write components accept arguments to override the local topic used for your own experimentation (please check out the main() functions to learn more). Since any topic can be passed in as an override, this is another reason to use the wildcard resource with the component permissions. Recall that this is okay for developing and testing, but the best practice for production components would be to exhaustively specify the permitted topics for publishing and subscribing.

The remainder of the recipe file is essentially the same, with differences simply in the naming of the component and the Python file to invoke in the Run script. Also, note that we add a new user group to ggc_user; the video group enables access to the LED matrix. Next, let's review the interesting new lines of code from this component's Python file.

Receiving messages on IPC

The business logic for receiving messages over IPC and writing messages to the LED matrix is coded in `scrolling_measurements.py`. Here are a few important sections to familiarize yourself with:

scrolling_measurements.py

```python
class StreamHandler(client.SubscribeToTopicStreamHandler):
    def __init__(self):
        super().__init__()


    def on_stream_event(self, event:
SubscriptionResponseMessage) -> None:
        try:
            message = event.json_message.message
            logger.info('message received! %s', message)
            scroll_message('t: ' + str("%.2f" %
message['temperature']))
            scroll_message('h: ' + str("%.2f" %
message['humidity']))
        except:
            traceback.print_exc()
```

In this selection, you can observe that a new local class is defined, called `StreamHandler`. The `StreamHandler` class is responsible for implementing the behavior of IPC client subscription methods such as the following:

- `on_stream_event` is the handler defining what to do when a new message arrives.

- `on_stream_error` is the handler defining what to do when the subscription encounters an error.

- `on_stream_close` is the handler defining how to clean up any resources when the subscription is closed.

Since the sensor component is publishing messages in JSON format, you can see that it is easy to get the value of that payload with `event.json_message.message`. Following this, the `on_stream_event` handler retrieves the values for both the `temperature` and `humidity` keys and sends a string to the `scroll_message` function. Here is the code for that function:

scrolling_measurements.py

```
def scroll_message(message):
    sense = SenseHat()
    sense.show_message(message)
```

That's it! You can view how easy it is to work with the Sense HAT library. The library provides the logic to manipulate the LED matrix into a scrolling wall of text. There are additional functions in the library for more fine-grained control of the LED matrix if scrolling a text message is too specific an action. You might want to render a solid color, a simple bitmap, or create an animation.

> **Note**
>
> In this pair of components, the messages transmitted over IPC use the JSON specification. This is a clean default for device software that can use JSON libraries since it reduces the code we have to write for serializing and deserializing messages over the wire. Additionally, using JSON payloads will reduce code for components that will exchange messages with the cloud via the AWS IoT Core service. This service also defaults to JSON payloads. However, both the IPC feature of IoT Greengrass and the AWS IoT Core service support sending messages with binary payloads.

In the context of the HBS hub device and appliance monitoring kit, the Raspberry Pi and its Sense HAT board are pulling double duty when it comes to representing both devices in our prototype model. It would be impractical to ask customers to review scrolling text on a screen attached to either device. In reality, the solution would only notify customers of an important event and not signal each time the measurements are taken. However, this pattern of sensor and actuator communicating through a decoupled interface of IPC topics and messages illustrates a core concept that we will use throughout the rest of the edge solutions built in this book.

Summary

In this chapter, you learned about the topologies that are common in building edge ML solutions and how they relate to the constraints and requirements informing architectural decisions. You learned about the common protocols used in exchanging messages throughout the edge and to the cloud, and why those protocols are used today.
You learned how to evaluate an edge ML solution for security risks and the best practices for mitigating those risks. Additionally, you delivered your first multi-component edge solution that maps sensor readings to an actuator using a decoupled interface.

Now that you understand the basics of building for the edge, the next chapter will introduce how to build and deploy for the edge using cloud services and a remote deployment tool. In addition to this, you will deploy your first ML component using a precompiled model.

Knowledge check

Before moving on to the next chapter, test your knowledge by answering these questions.

The answers can be found at the end of the book:

1. What are three network topologies that are common in edge solutions? Which one is implemented by the HBS hub device and appliance monitoring kit?

2. True or false: IoT Greengrass operates at the physical layer (that is, layer 1) of the OSI model.

3. What is the benefit of using a publish/subscribe model to exchange messages?

4. True or false: IoT Greengrass can act as both a messaging client and a messaging broker.

5. Is a message such as { "temperature": 70 } an example of structured data or unstructured data? Is it serializable?

6. Is image data captured from a camera an example of structured data or unstructured data? Is it serializable?

7. What do you think is the worst-case scenario if your home network router was compromised by an attacker but was still processing traffic as normal?

8. What is a mitigation strategy for verifying authenticity between two network devices?

9. Why is it important to protect root access via privilege escalation on a gateway device?

10. Is there any downside to wrapping every edge component in a container?

11. What functionality does IoT Greengrass provide to allow components to exchange messages?

12. What is one way to make the sensor and actuator solution you deployed in this chapter more secure? (Hint: review the recipe files!)

13. How might you redesign the sensor and actuator solution if you required a third component to interpret the sensor results before sending a message to the actuator?

References

Please refer to the following resources for additional information on the concepts discussed in this chapter:

- *The STRIDE Threat Model*: `https://docs.microsoft.com/en-us/previous-versions/commerce-server/ee823878(v=cs.20)?redirectedfrom=MSDN`

- *OSI model*: `https://en.wikipedia.org/wiki/OSI_model`

- *ISA95, Enterprise-Control System Integration*: `https://www.isa.org/standards-and-publications/isa-standards/isa-standards-committees/isa95`

- *PEP 405 -- Python Virtual Environments*: `https://www.python.org/dev/peps/pep-0405/`

- *Open Container Initiative*: `https://opencontainers.org/`

- Docker: `https://www.docker.com/`

- *AWS IoT Greengrass component recipe reference*: `https://docs.aws.amazon.com/greengrass/v2/developerguide/component-recipe-reference.html`

4

Extending the Cloud to the Edge

In the material leading up to this chapter, all of the development steps were performed on your device locally. Local development is useful for learning the tools and rapid prototyping but isn't representative of how you would typically operate a production device. In this chapter, you will treat your hub device as if it were actually deployed in the field and learn how to remotely interact with it using the cloud as a deployment engine.

Instead of authoring components on the device, you will learn how to use **Amazon Web Services (AWS) IoT Greengrass** to synchronize cloud resources, such as code, files, and **machine learning** (ML) models, to the edge and update devices via deployments. The tools, patterns, and skills you will learn in this chapter are important to your goals of extending the cloud to the edge and practicing how to manage an edge ML solution. You will connect a new client device to your hub device and learn how to bridge connectivity to a cloud solution. Finally, you will deploy your first real ML model to the edge. By the end of this chapter, you will be as familiar with pushing components and resources out to edge devices as you would for a production fleet.

In this chapter, we're going to cover the following main topics:

- Creating and deploying remotely
- Storing logs in the cloud
- Synchronizing the state between the edge and the cloud
- Deploying your first ML model

Technical requirements

To complete the hands-on steps for this chapter, you should have a hub device as defined by the hardware and software requirements in the Hands-on prerequisites section of *Chapter 1, Introduction to the Data-Driven Edge with Machine Learning*, and that device should be loaded with **AWS IoT Greengrass** core software, as defined by the steps from *Chapter 2, Foundations of Edge Workloads*. You will also need access to your **command-and-control** (**C2**) system, typically your PC or a laptop that has a web browser, and access to the AWS management console.

The resources provided to you for the steps in this chapter are available in the GitHub repository under the `chapter4` folder, at `https://github.com/PacktPublishing/Intelligent-Workloads-at-the-Edge/tree/main/chapter4`.

Creating and deploying remotely

Up until now, you have been interacting with your IoT Greengrass solution directly on the hub device using the IoT Greengrass **command-line interface** (**CLI**). Going forward, you will learn how to interact with your IoT Greengrass device from your C2 system (laptop or workstation) through the use of the AWS cloud. The CLI and local development lifecycle are great for learning the basics of IoT Greengrass and rapid iteration on component development. The best practice for production solutions is not to deploy the Greengrass CLI component to your devices and install new components locally, so next, you will learn how to package your components, store them in AWS, and complete remote deployments to your hub device.

Loading resources from the cloud

Components can include any number of artifacts defined by the recipe file, and those artifacts must be stored somewhere the Greengrass device can access them. Artifacts can be static resources of any kind: binary data such as images or text, application code such as Python source code or compiled **Java ARchive** files (**JARs**), containerized code such as a Docker container, and anything else that your component needs a copy of to work, the keywords here being **static** and **copy**. Artifacts are resources that your device uses that are copies that every other device deploying that component uses. They are also not intended to change on the device after deployment.

The other kind of resource that components use is **dynamic resources**. A dynamic resource gets loaded as a configuration specific to the device or something that gets consumed at runtime. Dynamic resources may be ephemeral in that they exist only as long as the device is online or a component is running. Some examples of dynamic resources are secrets to reference in your code such as an **application programming interface** (**API**) key, behavioral settings to include based on the device's **identifier** (**ID**) or other metadata, and communication channels between any leaf devices (those at terminal points in the solution, such as **Home Base Solutions** (**HBS**) appliance monitoring kits) and the Greengrass device (such as the HBS hub device), software components, and the cloud.

In the first hands-on section of this chapter, you will get familiar with how a recipe defines static resources as artifacts and perform your first remote deployment with a custom component. Later sections of this chapter will introduce implementations for dynamic resources.

Packaging your components for remote deployment

The main difference between local and remote deployment on IoT Greengrass is where the component resources come from. In your local development lifecycle, you were authoring files on the hub device itself and pointing to folders containing recipe and artifact files. For remote deployment, you must store your files in the AWS cloud using **Amazon Simple Storage Service** (**Amazon S3**) and update your recipes to point at S3 object locations. In the next steps, you will update the permissions model of your IoT Greengrass device to be able to read objects from S3, package up a component to S3 using the AWS CLI, and create a new remote deployment to your hub device.

Updating IoT Greengrass permissions

In your initial setup of IoT Greengrass on your hub device, a new role was created in the **AWS Identity and Access Management (AWS IAM)** service that the IoT Greengrass core service uses to get authorization to interact with any AWS services. In AWS, the AWS IAM service is where all users, groups, policies, and roles are defined that grant access to resources and APIs in your account.

A **user** is an IAM resource that serves as identification—for example, you as the **internet of things (IoT)** architect getting permissions to use APIs, or the `idtgg` user we defined for initially provisioning your hub device in AWS. A **policy** documents the sets of allowed and denied permissions that we attach to identities. A **role** is a façade of permissions that authorized users can assume to gain those specified permissions. A user gets permissions granted in one of three ways. Policies can be directly attached to the user or inherited from a group to which they belong. Policies can also be attached to roles that users assume for temporary sessions, letting the user perform specific session-based tasks following the **principle of least privilege (POLP)**. By using roles, we can define abstract sets of permissions that users can gain for the completion of specific tasks to follow the best practice of limiting permission scope. Here is a simple illustration of the relationship between a user (one type of security principal), a role, a policy, and the permissions granted to the user by assuming the role:

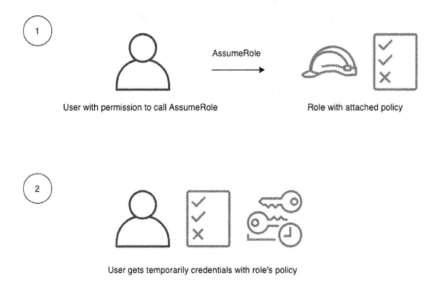

Figure 4.1 – An IAM user gets temporary permissions and credentials from a role

A real-world analogy for users, policies, and roles is when your friend temporarily gives you their house key while they are away so that you can water their plants or feed their cat. You can think of the role as *Someone Who Can Enter My House*, and the house key is the policy that grants access to their house. By trusting you with the key, your friend is granting you, the user, a temporary role to enter their home. When your task is complete and your friend returns home, you relinquish the key and thus end your session in that role. In the digital world, you as the user have your own key (such as a public certificate) that identifies you. The role and its policy grant your key temporary permission to access the resource instead of giving you the only key!

The installation process of IoT Greengrass created a new role named `GreengrassV2TokenExchangeRole` and attached to it a policy named `GreengrassV2TokenExchangeRoleAccess` that grants access to the APIs for interacting with the AWS IoT Core service and for writing logs to Amazon CloudWatch. By default, and as a best security practice, this policy does not include access to objects stored in S3. It is up to you as the solution developer to describe which S3 objects your devices should be able to access and add that configuration to the role as a new policy.

> **Note**
>
> For the rest of this chapter, steps with AWS CLI commands and simple filesystem management commands (such as the creation of new directories or files) will be written in Unix format for macOS and Linux systems. To get help using commands in AWS CLI where there are distinct differences for Windows, such as referencing local files as input, please refer to the AWS CLI documentation at `https://docs.aws.amazon.com/cli/latest/userguide/`.

In the next steps, you will use the AWS CLI to create a new bucket in S3, a new policy that grants read permissions to objects in that S3 bucket, and then attach the new policy to the role used by IoT Greengrass devices. Proceed as follows:

1. From your C2 system, open the terminal (or run `cmd.exe`/PowerShell on Windows).

2. Use the AWS CLI to print out your 12-digit account ID by running the following command: `aws sts get-caller-identity --query 'Account'` (the output should look like this: `012345678912`).

3. Create a new S3 bucket in your account. S3 bucket names share a global namespace in AWS. A good practice for naming buckets is to include your AWS account ID as a unique string. If a bucket name is already taken, you can add unique text (such as your initials) to the name until you find one that is not taken. Use the account ID from the previous step and replace the 12-digit account ID placeholder. Be sure to update the region if you're not using the default of `us-west-2`. The output should look like this: `aws s3 mb s3://012345678912-hbs-components --region us-west-2`.

4. From here on out, we will refer to this bucket name as `REPLACEME_HBS_COMPONENTS_BUCKET`. When you see a command or text in a file with the `REPLACEME_HBS_COMPONENTS_BUCKET` placeholder, you need to replace it with the name of your bucket, like this: `012345678912-hbs-components`.

5. Next, you will create a local file to store the content of the new policy that grants read access to the objects in your new bucket. You can find a template of this file in the GitHub repository for this book at the `chapter4/greengrass_read_s3.json` path and update the `REPLACEME_HBS_COMPONENTS_BUCKET` placeholder. If you're creating the file yourself, name it `greengrass_read_s3.json` and add the following content (remembering to replace the placeholder!):

greengrass_read_s3.json

```
{
    "Version": "2012-10-17",
    "Statement": [
        {
            "Effect": "Allow",
            "Action": [
                "s3:GetObject"
            ],
            "Resource": "arn:aws:s3:::REPLACEME_HBS_COMPONENTS_BUCKET/*"
        }
    ]
}
```

6. Create a new policy in IAM with this document as its source: `aws iam create-policy --policy-name GreengrassV2ReadComponentArtifacts --policy-document file://greengrass_read_s3.json`.

7. Add the policy to the IAM role used by IoT Greengrass. Replace the value of the `--policy-arn` argument with the `arn` value output from the previous command, as follows: `aws iam attach-role-policy --role-name GreengrassV2TokenExchangeRole --policy-arn arn:aws:iam::012345678912:policy/ GreengrassV2ReadComponentArtifacts`.

Now, your IoT Greengrass devices, such as the HBS hub device, can read component files that are stored in your S3 bucket. Let's cover one more element of how the security model works between the hub device and an AWS resource.

Earlier, we described the relationship in AWS IAM between a user, a role, and a policy. Your devices don't have any identity as a user in AWS IAM, so how do they assume the role? The answer is a feature of AWS IoT Core called the **Credentials Provider service**. Your devices do have an identity provisioned in AWS IoT Core using an **X.509** private key and public certificate. The credentials provider service is a means for connected devices to present their registered public certificates in exchange for temporary AWS credentials that have permissions granted through an IAM role.

Here is a sequence diagram showing the path for your HBS hub device to get permissions and ultimately read an object in S3:

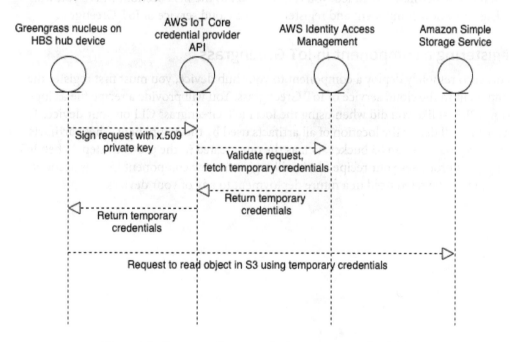

Figure 4.2 – Sequence diagram to fetch temporary credentials

As your edge solution adds features that interact with more AWS services, you must add permissions to those services using the same process you just completed. The same is true for your custom components that interact with AWS since all components get permissions through the IAM role of your device's configuration. As the number of distinct edge solutions in your account grows, the best practice is to create distinct IAM roles per distinct group of devices. For example, each HBS hub will have the same solution and can share a common role that defines permissions to access AWS. For the next project at HBS using IoT Greengrass, instead of adding more permissions to the same IAM role, it is best to create a new role for devices of that project.

> **Note**
>
> There is a regional quota for the number of roles you can define for your devices in AWS IoT. As of this writing, the quota is 100. This means you don't want to create a unique role per Greengrass device, as you will quickly reach the maximum quota as you scale. A best practice for a production solution on AWS is to use one AWS account per production solution, such as maintaining the fleets of hub devices at HBS. A different product line may be deployed in a separate AWS account, thus expanding the total number of roles to use.

With the new permissions added, you can move on to the next section, where you will package up a new component and register it in the cloud service of IoT Greengrass.

Registering a component in IoT Greengrass

In order to remotely deploy a component to your hub device, you must first register the component in the cloud service of IoT Greengrass. You will provide a recipe file as input to an API, just like you did when using the local IoT Greengrass CLI on your device. The recipe file will define the location of all artifacts used by the component. These artifacts must be uploaded to an S3 bucket such as the one created in the previous step. When IoT Greengrass processes your recipe file, it will register a new component in your account that can then be referenced in a future deployment to any of your devices.

The first step in registering a component is to add your artifact files to the S3 bucket. Only then will you know the addresses of those files in S3 to be able to update a recipe that can reference them. In this section, you will upload a ZIP archive as an artifact, replace the path to the artifact in the recipe file, then register the component in your account using the AWS **software development kit (SDK),** as follows:

1. From the book's GitHub repository, change directory to `chapter 4` by running the following command: `cd chapter4`.

2. Use the AWS SDK to upload the artifact file by running the following command: `aws s3 cp artifacts/com.hbs.hub.HelloWithConfig/1.0.0/ archive.zip s3://REPLACEME_HBS_COMPONENTS_BUCKET/ artifacts/com.hbs.hub.HelloWithConfig/1.0.0/archive.zip`.

3. Edit the `recipes/com.hbs.hub.HelloWithConfig-1.0.0.json` file and replace the value of the **Uniform Resource Identifier (URI)** key with the path to your artifact in S3 (the last argument in the previous step). After filling it in, it should look something like this:

    ```
    "Artifacts": [
        {
            "URI": "s3://012345678912-hbs-components/artifacts/
    com.hbs.hub.HelloWithConfig/1.0.0/archive.zip",
            "Unarchive": "ZIP",
    ```

4. Now that your artifact is in S3 and your recipe file is updated, you can use the AWS SDK to register your new component in the cloud service of IoT Greengrass. In the response will be the **Amazon Resource Name (ARN)** of your component that you will use for future steps. Run the following command: `aws greengrassv2 create-component-version --inline-recipe fileb://recipes/ com.hbs.hub.HelloWithConfig-1.0.0.json`.

5. The preceding command returns a status such as `componentState=REQUESTED` to signal that IoT Greengrass is taking steps to register your new component. To check on the status of your component registration, run the following command (replacing the `--arn` argument with the one found in the output from the previous step): `aws greengrassv2 describe-component --arn arn:aws:greengrass:us-west-2:012345678912:components:com. hbs.hub.HelloWithConfig:versions:1.0.0`.

When the component is registered in the service, you will see a response to this command with the `componentState` value now showing as DEPLOYABLE. This means the component is now ready for inclusion in a new deployment to a device. Before moving on to the deployment, let's take a look at the recipe file now stored by IoT Greengrass with the following command (replace the `--arn` argument with your component's ARN from previous steps): aws greengrassv2 get-component --arn arn:aws:greengrass:us-west-2: 012345678912:components:com.hbs. hub.HelloWithConfig:versions:1.0.0 --query "recipe" --output text | base64 --decode. You may notice the recipe file doesn't look exactly like the one you sent to IoT Greengrass. Here's what the `Artifacts` object looks like now:

```
"Artifacts":[{"Uri":"s3://012345678912-hbs-components/
artifacts/com.hbs.hub.HelloWithConfig/1.0.0/archive.
zip","Digest":"wvcSArajPd+Ug3xCdt0P1J74/I7QA2UbuRJeF5ZJ7ks=","A
lgorithm":"SHA-256","Unarchive":"ZIP","Permission":{"Read":"OWN
ER","Execute":"OWNER"}}]
```

What's new here are the `Digest` and `Algorithm` keys. This is a security feature of IoT Greengrass. When your recipe is registered as a component in the service, IoT Greengrass computes a `SHA-256` hash of each artifact file referenced by the recipe. The purpose is to ensure that the artifacts eventually downloaded by any IoT Greengrass devices have not been modified before use. This also means you cannot alter an artifact file stored on S3 after registering the component. To update any artifact requires you to register a new version of the component and deploy the new component version.

Two more deltas between this component and components developed locally in previous chapters are the use of the decompressed path and artifact permissions. Here is a snapshot of the key differences in this recipe file:

```
"Lifecycle": {
  "Run": "cd {artifacts:decompressedPath}/archive && ./hello.
sh"
},
"Artifacts": [
  {
    "URI": "s3://REPLACEME_HBS_COMPONENTS_BUCKET/artifacts/com.
hbs.hub.HelloWithConfig/1.0.0/archive.zip",
    "Unarchive": "ZIP",
    "Permission": {
      "Execute": "OWNER"
```

In the `Lifecycle` object, you can see the run script that makes reference to the `artifacts:decompressedPath` variable. This variable points to the directory where Greengrass automatically unarchives your archived artifacts. Files are unpacked to a subdirectory with the same name as the archive—in this case, `archive/`. We know our `hello.sh` script will reference the adjacent `config.txt` file from the same archive. We must tell the run script to change directory to the decompressed path and then run the script in order to find the `config.txt` file in the correct directory context.

The best practice for your artifacts is to consume them from the `artifacts` directory, as downloaded or unpacked by IoT Greengrass, and to use the component's work directory, available in the recipe as `work:path`, for files to which your component will write data. The `work` directory is the default context for any lifecycle script, and that is why we include a `change` directory command before running our script artifact.

The other new inclusion is the `Permission` object, where you can see we are setting an `Execute` property to the `OWNER` value. By default, artifacts and files from unpacked artifacts have a filesystem permission of `Read` for the component's owner (such as the default `ggc_user`). This means a script file in our `archive.zip` file would not be executable without a change to the file's permissions. Using the `Permission` object for any artifact, we can set the `Read` and `Execute` filesystem permissions to any of `NONE`, `OWNER`, or `ALL` (all system users). This is also related to why artifacts are write-protected. Artifacts are intended to be read-only resources consumed by the component or executable files that should not be changed without a revision to the component's definition.

In the next section, you will deploy your newly registered component to your device.

Remotely deploying a component

With your component now available in the IoT Greengrass service, it's time to start a remote deployment from your C2 system to your HBS hub device using the AWS CLI. This kind of deployment from a remote system using the IoT Greengrass cloud service is the standard way by which you will deploy updates to your devices. It may not always be a manual step from your developer laptop, but we will cover more details about the production pipeline in *Chapter 8, DevOps and MLOps for the Edge*.

In the local development lifecycle, the only device that you were deploying your component to was the local one. Deployment through the cloud service of IoT Greengrass is how you specify multiple target devices. These kinds of deployments are how you scale up your ability to manage a fleet of any size that should all have the same components running on them. A deployment will also specify rollout and success criteria, such as the rate at which to notify devices, how long they have to report a successful deployment, how long to wait for components to signal they are ready for updates, and what to do if the deployment fails.

In the following steps, you will write a file that tells IoT Greengrass about the details of your deployment, initiate the deployment, and then verify the deployment's success:

1. You will need the ARN of the **thing group** to which your HBS hub device belongs. A thing group is an addressing mechanism of AWS IoT Core for grouping like devices together. You created the `hbshubprototypes` thing group as an argument in the initial IoT Greengrass core software installation. The following command will fetch the ARN of your thing group that you will use in the next step: `aws iot describe-thing-group --thing-group-name hbshubprototypes --query "thingGroupArn"`.

2. Edit the `chapter4/deployment-hellowithconfig.json` file from the GitHub repository and replace the value of `targetArn` with the thing group ARN output from the previous step. After editing the file, it should look something like this:

```
{
    "targetArn": "arn:aws:iot:us-west-
2:012345678912:thinggroup/hbshubprototypes",
    "components": {
      "aws.greengrass.Cli": {
        "componentVersion": "2.4.0"
      },
      "com.hbs.hub.HelloWithConfig": {
        "componentVersion": "1.0.0"
      }
    }
}
```

3. Start a new deployment to the group containing your hub device using the following deployment configuration: `aws greengrassv2 create-deployment --cli-input-json file://deployment-hellowithconfig.json`.

4. The previous command starts the deployment process. To check on the status of the deployment on a particular device, such as your `hbshub001` device, you can use the following command: `aws greengrassv2 list-effective-deployments --core-device-thing-name hbshub001`.

5. To validate on your device that the component ran as intended, you can log in or use a **Secure Shell** (**SSH**) back into your device and check the logs with `sudo less /greengrass/v2/logs/com.hbs.hub.HelloWithConfig.log` as follows:

```
2021-07-27T20:19:07.652Z [INFO] (pool-2-thread-91)
com.hbs.hub.HelloWithConfig: shell-runner-start.
{scriptName=services.com.hbs.hub.HelloWithConfig.
lifecycle.Run, serviceName=com.hbs.hub.HelloWithConfig,
currentState=STARTING, command=["./hello.sh"]}

2021-07-27T20:19:07.685Z [INFO] (Copier) com.hbs.
hub.HelloWithConfig: stdout. Hello from Zimbabwe!.
{scriptName=services.com.hbs.hub.HelloWithConfig.
lifecycle.Run, serviceName=com.hbs.hub.HelloWithConfig,
currentState=RUNNING}

2021-07-27T20:19:07.699Z [INFO] (Copier) com.
hbs.hub.HelloWithConfig: Run script exited.
{exitCode=0, serviceName=com.hbs.hub.HelloWithConfig,
currentState=RUNNING}
```

At this point, you have completed your first remote deployment to your hub device using IoT Greengrass! The overall process is not too different from local development. We needed to upload our artifacts to a cloud service such as S3, update the recipe file to point to these new artifact locations, and register the component before including it in a deployment. The deployment itself also has a few more options for specifying which devices to target, behavior for scaling out to a fleet, and criteria and behavior of success or failure.

Each thing group has a 1:1 mapping with a deployment that represents the latest configuration each device in that group should be using. When you want to deploy a change to a group of devices, you will create a revision to the deployment instead of starting an all-new deployment. A revision still takes a deployment configuration similar to the one we used in this section and expects an explicit definition of all components and configuration, meaning it is not an amendment to the last known deployment.

You can include a device in multiple thing groups, where each thing group defines a deployment of unique configuration and components. For example, you could define a thing group of monitoring components that get applied to all HBS devices, and pair this with thing groups that specify business logic components based on the kind of device it is, such as our smart home hub. A device that belongs to multiple thing groups will receive deployment notifications for each group and merge the component graph from all of them. Here is an illustration of how we can use thing groups to effectively manage the components that get deployed across a fleet of devices to build up an aggregate solution:

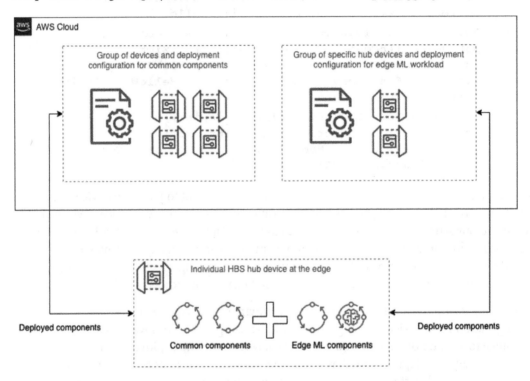

Figure 4.3 – Example of aggregating components across group deployments

As mentioned earlier, your deployment configuration can specify whether to allow components an option to signal they are ready for a restart or update. The value of letting components interact with deployment behavior in this way is to prevent any loss of data or business process interruption from the component suddenly being terminated. The component must include the use of IoT Greengrass **interprocess communication** (**IPC**) with the IoT Greengrass SDK and implement the `SubscribeToComponentUpdates` function.

A component can then respond to component update events and request a deferment by publishing a message back over IPC using the `DeferComponentUpdate` command. There is a similar operation for components to validate configuration change requests with `SubscribeToValidateConfigurationUpdates` and the respective `SendConfigurationValidityReport` features. You can learn more about these features from the *References* section at the end of this chapter.

With your first remote deployment complete and with a better understanding of how the deployment service of IoT Greengrass works, let's make it easier to remotely troubleshoot your hub device and its components by enabling the publication of local logs to the cloud.

Storing logs in the cloud

An edge solution loads resources from the cloud in order to bootstrap, configure, and generally make the local solution ready for runtime. The health of the device and the solution should also be reported to the cloud to assist with production operations. By default, your HBS hub device checks in with the cloud IoT Greengrass service to report connectivity status and the result of the most recent deployment. To get more telemetry from the edge solution, such as logs and metrics, you need to deploy additional components. In this section, you will deploy a managed component that ships component logs up to Amazon CloudWatch, a service for storing and querying logs.

Storing the logs of our hub devices in the cloud is a best practice and enables us to triage devices individually without needing a live connection to the device or being physically in front of it. After all, some of these devices may be in rather remote locations such as the Alaskan tundra or only come online at scheduled times, such as the narwhal-studying submersible from *Chapter 1, Introduction to the Data-Driven Edge with Machine Learning*.

Another benefit of storing logs in the cloud is that you can run queries across a fleet of devices to discover insights about fleet performance. For example, a simplistic count of lines per device log over a 24-hour period could find outliers of *chatty* devices where there is an abnormal amount of activity, which could mean the device is processing an unusual amount of data or thrashing resources. The following histogram of log activity across our fleet would indicate a potential problem for your operations team to triage:

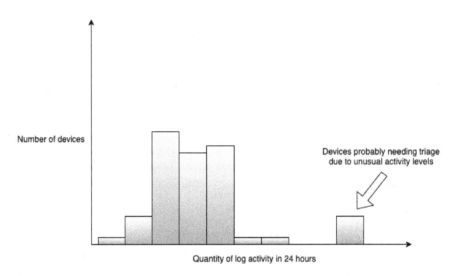

Figure 4.4 – Sample histogram showing device outliers based on log activity

Since storing logs is such a common use case for an edge solution and also in the wider spectrum of cloud application development, IoT Greengrass provides a managed component to make it easy to ingest your components' logs. This managed component, called aws.greengrass.LogManager, is authored and maintained by AWS. Managed components are anticipated as fulfilling common requirements of IoT architects but deemed as opt-in features that you need to bring in with your deployments.

> **Note**
>
> The latest version of the aws.greengrass.LogManager managed component at the time of this writing was *2.2.0*. You may need to update the version based on the latest version of IoT Greengrass core software installed when you started this book. For the steps in *Chapter 2, Foundations of Edge Workloads*, IoT Greengrass core software version *2.4.0* was used, which is compatible with LogManager *2.2.0*. You can see the latest dependency information in the *AWS-provided components* documentation link in the *References* section found at the end of this chapter.

In the following steps, you will revise the deployment for the `hbshubprototypes` group to include the `aws.greengrass.LogManager` managed component. The configuration for the `LogManager` component will specify which components to upload log files to. Then, you will use the AWS CLI to run a simple query to validate that log files are being stored. Proceed as follows:

1. Edit the `chapter4/deployment-logmanager.json` file to swap the placeholder with your account ID. This deployment adds the `LogManager` component. Remember that by not specifying other components in the deployment list—such as `com.hbs.hub.HelloWithConfig`, which you added in a previous section—they will be removed from the device. We will remove `HelloWithConfig` for this deployment so that we can see the runtime output to the log file when we add it back. All you need to do is update the `targetArn` property to replace the account ID placeholder and save the file.

2. Create a new deployment revision and pass in this new deployment configuration file, as follows: `aws greengrassv2 create-deployment --cli-input-json file://deployment-logmanager.json`.

3. Edit the `chapter4/deployment-logmanager.json` file to once again add the `HelloWithConfig` component. We do this to redeploy the component so that the runtime output is written to the log again and will then be uploaded to the cloud. Add the following bolded lines into the `components` object and save the file:

```
{
    "targetArn": "arn:aws:iot:us-west-
2:0123456789012:thinggroup/hbshubprototypes",
    "components": {
      "aws.greengrass.Cli": {
        "componentVersion": "2.4.0"
      },
      "com.hbs.hub.HelloWithConfig": {
        "componentVersion": "1.0.0"
      },
      "aws.greengrass.LogManager": {
```

4. Create a new deployment revision using the same command as before, like this: `aws greengrassv2 create-deployment --cli-input-json file://deployment-logmanager.json`.

5. Once this deployment is complete, you will start seeing logs in CloudWatch Logs within 5 minutes since the configuration for `LogManager` specifies `300` seconds as the `periodicUploadIntervalSec` parameter.

6. Use the following command to check on the status of new log groups with the prefix: `/aws/greengrass/`: `aws logs describe-log-groups --log-group-name-prefix "/aws/greengrass/"`.

7. You know logs are being written to CloudWatch when you see a response such as the following:

```
{
    "logGroups": [
        {
            "logGroupName": "/aws/greengrass/
UserComponent/us-west-2/com.hbs.hub.HelloWithConfig",
            "creationTime": 1627593843664,
            "metricFilterCount": 0,
            "arn": "arn:aws:logs:us-west-
2:012345678912:log-group:/aws/greengrass/UserComponent/
us-west-2/com.hbs.hub.HelloWithConfig:*",
            "storedBytes": 2219
        }
    ]
}
```

8. You can query the log group to see devices that are storing logs for this component, as follows: `aws logs describe-log-streams --log-group-name /aws/greengrass/UserComponent/us-west-2/com.hbs.hub.HelloWithConfig`.

9. The response, as illustrated in the following code snippet, will show log streams named in a `/DATE/thing/THING_NAME` format:

```
{
    "logStreams": [
        {
            "logStreamName": "/2021/07/29/thing/hbshub001",
```

10. Plug the log group name into the `filter-log-events` command to see the logged output of the `HelloWithConfig` component, as follows: `aws logs filter-log-events --log-group-name /aws/greengrass/ UserComponent/us-west-2/com.hbs.hub.HelloWithConfig --filter-pattern stdout`.

The output is as follows:

```
{
    "events": [
        {
            "logStreamName": "/2021/07/29/thing/
hbshub001",
            "timestamp": 1627579655321,
            "message": "2021-07-29T17:27:35.321Z [INFO]
(Copier) com.hbs.hub.HelloWithConfig: stdout. Hello
from Zimbabwe!. {scriptName=services.com.hbs.hub.
HelloWithConfig.lifecycle.Run, serviceName=com.hbs.hub.
HelloWithConfig, currentState=RUNNING}",
            "ingestionTime": 1627593843867,
            "eventId": "362…"
        }
    ]
}
```

You can see from this series of instructions how to include the `LogManager` component that will automatically ship your components' log files to the cloud and how to use the Amazon CloudWatch Logs service to query into your logs. IoT Greengrass makes it easy to triage log files of one component across a fleet of devices by querying the log group or into individual devices by querying the log stream that represents one device's component. For more powerful log analysis tooling, you can explore Amazon CloudWatch Logs Insights for aggregation queries across log groups, or stream logs into an indexed querying tool such as **Amazon Elasticsearch**.

Merging component configuration

Now is a good opportunity to introduce how to merge in component configuration at deployment time, which you can see we are doing in deployment with `logmanager.json`. If you recall from earlier recipe creation in the *Writing your first component* section of *Chapter 2, Foundations of Edge Workloads*, the recipe can define a `ComponentConfiguration` object that specifies the default settings that a component will use at runtime. We used this to define a default `World!` text that gets passed into the `HelloWorld` component. This kind of component configuration can also be defined at the time of deployment to one or more devices to override these defaults. This is useful when setting the configuration for distinct groups of devices, such as telling all of our prototype devices to use verbose debug-level logging and our production devices to use warning-level logging to save on costs. Here is an illustration of that practice in effect:

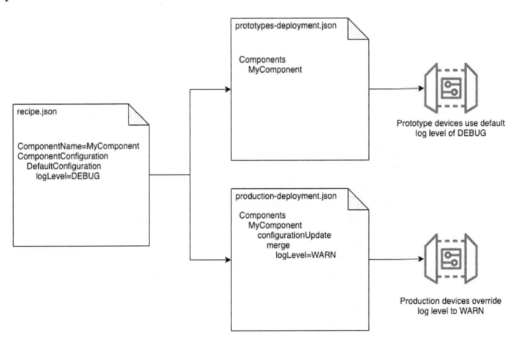

Figure 4.5 – Overriding configuration for fleets of devices

A deployment can define component configuration using either of two properties, `reset` and `merge`. The `reset` property tells the component to restore the configuration to whatever the default is for the given configuration key. The `merge` property tells the component to apply a new configuration for the given configuration key, without affecting existing values of other configuration keys. Using both the `reset` and `merge` property on the same configuration key will always first reset the value and then merge in the new value. This can be useful for restoring a tree of default values and then merging in an update for just one node in the tree.

If you inspect the `deployment-logmanager.json` file, you can see the deployment-time configuration merge we are using to tell the `LogManager` component what to do. Here is a pretty-print version of the `merge` object:

```json
{
  "logsUploaderConfiguration": {
    "systemLogsConfiguration": {
      "uploadToCloudWatch": "true",
      "minimumLogLevel": "INFO",
      "diskSpaceLimit": "10",
      "diskSpaceLimitUnit": "MB",
      "deleteLogFileAfterCloudUpload": "false"
    },
    "componentLogsConfigurationMap": {
      "com.hbs.hub.HelloWithConfig": {
        "minimumLogLevel": "INFO",
        "diskSpaceLimit": "10",
        "diskSpaceLimitUnit": "KB",
        "deleteLogFileAfterCloudUpload": "false"
      }
    }
  },
  "periodicUploadIntervalSec": "300"
}
```

Without any merged configuration set in the deployment, the version 2.2.0 `LogManager` component used here doesn't actually do anything. You must give it some configuration at deployment time to get any logs sent to the cloud.

In the preceding sample, there is a `logsUploaderConfiguration` configuration key that has two child nodes and an interval property. The `systemLogsConfiguration` node tells the `LogManager` component to upload to Amazon CloudWatch IoT Greengrass system logs such as `greengrass.log`. The `componentLogsConfigurationMap` node tells the `LogManager` component how to selectively upload logs for your other components. You can see here we are defining a `com.hbs.hub.HelloWithConfig` component for inclusion to send logs to the cloud. You would add one object to this list for each component to explicitly capture logs. Two best practices to consider are outlined here:

- Generate your deployment configuration programmatically and build out the `LogManager` configuration based on the other components included in that deployment. A script in a build process that inspects the components included in your deployment can update the `LogManager` `componentLogsConfigurationMap` node before it gets passed to the `CreateDeployment` API.

- Create a thing group, such as `CommonMonitoringTools`, put all of your Greengrass devices in it, and set a group-level deployment configuration to capture the system logs in the `systemLogsConfiguration` node. All of your devices would then include this component and configuration, resulting in a default behavior to upload system logs. A separate thing group deployment that represents your application's components would then merge the `LogManager` configuration for the `componentLogsConfigurationMap` node in order to specify the logs for that application's components. This works because two deployments from two different thing groups can stack on a single device, effectively merging the configuration of a single component. *Figure 4.3* and *Figure 4.5* together illustrate this concept.

One last note on configuration management is addressing the delta between the preceding pretty-printed **JavaScript Object Notation (JSON)** and the escaped JSON string you see in `deployment-logmanager.json`. At the time of this writing, the IoT Greengrass deployment API only accepts the configuration as a string object, so the configuration must be defined as JSON and then escaped as a single string before sending it to the deployment API. This is more inconvenient when hand-writing deployment files but is a simple added step when building deployments programmatically. An alternative, in this case, could be to define your deployment files using the **YAML Ain't Markup Language (YAML)** format instead of JSON because the YAML specification has syntactical support for constructing multiline inputs.

You now have a functioning, managed component for storing your hub's log files in the cloud to facilitate remote diagnostics for your edge solution. You know how to add more components to the `LogManager` component by merging in new changes to the configuration. Through that process, you learned more about the component configuration system of IoT Greengrass that will serve you as you create new components and build deployments with multiple components working together. In the next section, you will learn how leaf devices connected to your hub can exchange messages and synchronize the state with the cloud.

Synchronizing the state between the edge and the cloud

The HBS hub device, if made into a real product, would connect with local devices over a network protocol and proxy telemetry and commands with a cloud service. In the previous chapter, we used components running on our hub device to interface with local hardware interfaces on the **Raspberry Pi Sense HAT**.

This makes sense when the hub device communicates with hardware over serial interfaces, but when communicating over a network, those appliance monitoring kits won't really be software components running on the hub device using the **Greengrass IPC** interface to exchange messages. Instead, they may use a network protocol such as **Message Queue Telemetry Transport (MQTT)** to exchange messages with the hub device over Wi-Fi or Bluetooth.

In this section, you will deploy new managed components for connecting to leaf devices over MQTT and synchronize the state of a leaf device's telemetry to the cloud.

Introduction to device shadows

By synchronizing state, we mean the end result of keeping two systems up to date with the latest value. When working with the edge and cloud, we must acknowledge and work with the reality that edge solutions may not currently be connected to the cloud service. Our leaf devices and the hub device may work with acquired telemetry that has yet to be reported to the cloud. Once the hub device restores the connection with the cloud, there must be a mechanism for reconciling the current state of the edge and the current state of the cloud.

For example, if our hub device is disconnected from the cloud because of a network drop, new telemetry will be acquired by the leaf devices and new remote commands could be queued up from the cloud service from the customer's mobile application. When the connection is restored, something needs to update the device state so that everything gets back in sync. For this purpose, AWS IoT Core offers a service called **Device Shadow** that acts as a synchronization mechanism for devices and the cloud. IoT Greengrass makes this Device Shadow service available at the edge via a managed component.

A device shadow is a JSON document that summarizes the current state of reported data and the desired state. Typically, this means that the device is responsible for updating the reported data, and other actors in the system instruct the device using the desired state. Let's say our hub device is supposed to keep the cloud informed of the latest temperature measurement from one of the appliance monitoring kits. Let's also say that the customer's mobile application can send a command to restart the monitoring kit as a form of troubleshooting. The following diagram illustrates how these messages are stored in the device shadow and reconciled after resuming from a network drop event:

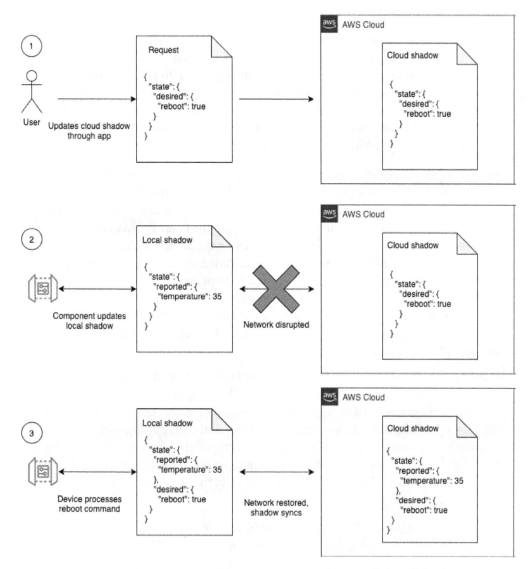

Figure 4.6 – Flow of shadow messages synchronizing after a network disruption

Device shadows are also useful for our edge ML workloads since a component running an inference task can subscribe to changes reported by the shadow service, or even register its own shadow for exchanging state and commands with the cloud and other components running on the Greengrass device.

In the *Connecting your first device: sensing at the edge* and *Connecting your second device: actuating at the edge* sections of *Chapter 3, Building the Edge*, you implemented IPC in order to let components exchange information in real time. Shadows can level up your edge solutions by defining synchronized state documents that also get communicated over IPC or even synchronized with the cloud. This opens up some interesting use cases that further let you as the architect focus on solving business problems without also engineering mechanisms for data exchange. Here are a few examples of how ML workloads running as edge components can benefit from using shadows:

- **Reporting and caching application state**: Your ML inference component can create a local shadow that stores the latest inference results whenever the ML model is used to process new data. If the model is trained to report when there is an anomaly detected on new data, the local shadow could store the latest anomaly score and confidence rating. That way, other components on the device can subscribe to changes in the shadow and alter their behavior based on the output of the ML component. This may seem similar to publishing updates over IPC, but the key difference here is that components can get values from the shadow whether or not they were running at the same time as when the ML component last published the anomaly score. In that sense, the shadow is used as a caching mechanism and helps us decouple our edge solution.

- **Sending commands to components**: Shadows can be used to instruct components of new commands or desired behavior, even if the component is not currently running. For example, if a component crashes or is in some other recovery state at the time a command would have otherwise been sent to it, putting that command in a shadow ensures it will be delivered to the component when it next enters its running state. Combined with synchronizing shadows to the cloud AWS IoT Core service, this enables other applications and devices to interact with components at the edge in a resilient way.

Now that we have introduced state synchronization with shadows, let's move on to the hands-on steps for deploying this functionality to your solution.

Steps to deploy components for state synchronization

IoT Greengrass provides the functionality for connecting leaf devices, storing shadows, and synchronizing messages to the cloud with a few separate managed components. Those components are an **MQTT broker** (`aws.greengrass.clientdevices.mqtt.Moquette`), **Shadow Manager** (`aws.greengrass.ShadowManager`), and an **MQTT bridge** (`aws.greengrass.clientdevices.mqtt.Bridge`), respectively.

The following diagram illustrates the architecture of this solution and how these components work together to deliver state synchronization for devices:

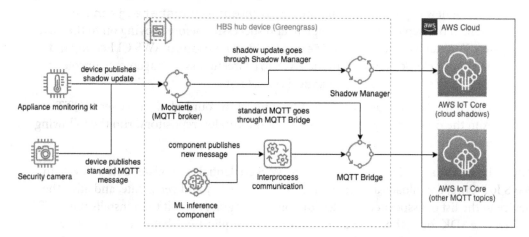

Figure 4.7 – Examples of message flows using managed components

By following the next steps of this section, you will create a new deployment that enables all three of these managed components to launch a new standalone application on your C2 system that connects to your hub device over MQTT and observe how the state is synchronized between the edge and the cloud using these features. Proceed as follows:

1. Update your Greengrass role in AWS IAM to allow the use of the device shadow APIs of AWS IoT Core. These steps are specific to AWS IAM and represent a common flow of preparing your Greengrass device to use new APIs. Then, do the following:

    A. From the `chapter4` directory, create a new IAM policy, as follows: `aws iam create-policy --policy-name greengrass-use-shadows --policy-document file://greengrass_use_shadows.json --description "Allows Greengrass cores to interact with AWS IoT Core device shadows"`.

    B. Copy the policy ARN from the response of the `create-policy` command and use it in the `attach-role-policy` command next. It will look like this, but with your account number instead: `arn:aws:iam::01234567890:policy/Greengrass-use-shadows`.

    C. Attach the new policy to your Greengrass IAM role, replacing the `policy-arn` argument with your own from the previous step, as follows: `aws iam attach-role-policy --role-name GreengrassV2TokenExchangeRole --policy-arn arn:aws:iam::01234567890:policy/Greengrass-use-shadows`.

2. Edit the `deployment-syncstate.json` file to update the `targetArn` property for your account ID. Save the file.

3. Create a new deployment using the `deployment-syncstate.json` file. Confirm the deployment is complete and successful before moving on to the next step. You can use the `list-effective-deployments` AWS CLI command to verify or check the status in the AWS IoT Greengrass management console at `https://console.aws.amazon.com/iot`.

4. Launch the `local device` client to connect to your hub device over MQTT. From the `chapter4/apps` directory of the GitHub repository, run the following command: `./client.sh`.

The script at `apps/client.sh` uses your AWS credentials to register a new device in AWS IoT Core, downloads a generated `x.509` private key and certificate, and adds the device to the list of associated devices for your Greengrass core. It then installs the AWS IoT Device SDK for Python and launches a Python application that discovers your hub device on the local network and connects to it for exchanging messages over MQTT. The purpose of running this application off of the hub device is to demonstrate how Greengrass supports exchanging messages for local devices.

> **Note**
>
> If your C2 laptop is not on the same network as your hub device, you can run the client application on your hub device. However, you will need to update the permissions for the `idtgg` AWS user that was configured on the hub device since it does not have the necessary permissions yet. Return to the AWS management console for IAM, find the user group named `Provision-IDT-Greengrass`, add a new permission by attaching a policy, and choose the policy named `AWSIoTFullAccess`. Then, you can run the `client.sh` application on your hub device using the existing credentials there from *Chapter 2, Foundations of Edge Workloads*.

The `client.sh` application will publish new messages on a topic dedicated to device shadow communications. The topic address is `$aws/things/localdevice/shadow/update`. The client reports a randomly generated temperature value as if it were another sensor on the local network. You can verify that these messages are making it up to the cloud by using the AWS IoT Core management console at `https://console.aws.amazon.com/iot`. Navigate to **Test > MQTT test client** and subscribe to the `$aws/things/localdevice/shadow/update/accepted` topic to see the result of the publishing. Here is an example of what it looks like:

$aws/things/localdevice/shadow/update/accepted

| Pause | Clear | Export | Edit |

▼ $aws/things/localdevice/shadow/update/accepted

September 25, 2021, 17:53:49 (UTC-0600)

```
{
  "state": {
    "reported": {
      "temperature": 32
    }
  },
  "metadata": {
    "reported": {
      "temperature": {
        "timestamp": 1632614028
      }
    }
  },
  "version": 1,
  "timestamp": 1632614028
}
```

Figure 4.8 – Screenshot from AWS IoT Core console of the accepted shadow update

You can interact with the local device application through this test client, too. You can send a new desired state to the local device and it will merge the command into the next reported message. Use the MQTT test client to publish a message on the $aws/things/localdevice/shadow/update topic with this body: {"state":{"desired":{"valve":"open"}}}. You should see on your subscription and in the standard output of the client.sh application activity that it shows the open valve now in the reported state instead of the desired state. This means the local device received a command through Greengrass, processed the command, and updated its reported state accordingly.

You can continue experimenting with this by publishing further messages to the desired state through the MQTT test client. A real device would take some action in response to the desired state, but this application just copies the command from the desired state to the reported state to simulate handling the event.

To demonstrate the functionality of the MQTT bridge managed component, the `client.py` application is also publishing a heartbeat message on another topic at the `dt/localdevice/heartbeat` address. The reason for this is that the Shadow Manager component actually handles all the synchronization of state for us when using the shadow topics. For any other topics and data published, such as this heartbeat behavior, we must use the MQTT bridge component to ferry messages from the edge to the cloud (or vice versa). Use the MQTT test client to subscribe on `dt/localdevice/ heartbeat` to see these messages arrive in the cloud.

Something fun to try is to take your hub device offline temporarily and repeat the previous step but set the desired state of the valve to `closed`. This will demonstrate how the shadow service messages are buffered for delivery the next time the Greengrass device comes back online. On your hub device, bring down the network connection (either by unplugging the Ethernet cable or disabling Wi-Fi). You should see no more messages arriving on either the shadow or heartbeat topics in the MQTT test client. Publish a new message in the MQTT test client, such as `{"state":{"desired":{"valve":"open"}}}`, and then bring the network connection back up. You will see the local device get the new command once the hub device reestablishes connectivity as well as a resumption of the heartbeat messages.

Extending the managed components

The deployed example uses components managed by AWS to establish connectivity for local devices over an MQTT broker, sending and receiving messages between the edge and the cloud, and synchronizing state with the device's shadow service. You can configure these components for your own use cases by merging further configuration updates. Let's examine the configuration used in this deployment, then review a few more options and best practices.

> **Note**
>
> For any configuration JSON reviewed in this section, the book is using a pretty-print format that doesn't match what you see in the actual deployment files where JSON is *stringified* and escaped per the API requirements. This is done for convenience and legibility in the book format.

Shadow Manager

In the `aws.greengrass.ShadowManager` component configuration of the `deployment-syncstate.json` file, we must explicitly define which device shadows will be kept in sync between the edge and cloud. The configuration we used defined the shadow to synchronize for a device named `localdevice` that matches the client ID of the `apps/client.sh` application. Here is the configuration used to achieve that:

```
{
    "synchronize":{
      "shadowDocuments":[
        {
           "thingName":"localdevice",
           "classic":true,
           "namedShadows":[]
        }
      ]
    }
}
```

A **classic shadow** is a root-level shadow for any device in AWS IoT Core. **Named shadows** are children that belong to the device, and a device can have any number of additional children shadows to help with **separation of concerns (SoC)**. In this case, just using a classic shadow is sufficient.

For each device you want your Greengrass solution to keep in sync, you would need the configuration to specify that device in the list of `shadowDocuments` where `thingName` is the name of your device that it uses to connect over MQTT (AWS IoT Core defines many resources and APIs representing devices as *things* per the IoT moniker).

Your Greengrass core device is itself a thing registered in AWS IoT Core and can use both classic and named shadows. These shadows can be used for representing the state of the hub device itself or for components on the hub device to behave in coordination with some desired command; for example, sending the desired state to the hub device's shadow to enter a low bandwidth state could get picked up by any components running on the device to act accordingly. Enabling shadow synchronization for your Greengrass core device has a separate configuration outside of the `shadowDocuments` property. Here is an example of how that could look:

```
{
    "synchronize":{
      "coreThing":{
```

```
        "classic":true,
        "namedShadows":[
          "CoreBandwidthSettings",
          "CorePowerConsumptionSettings"
        ]
    },
    "shadowDocuments":[ ... ]
}
```

This section covered what you need to know about shadows in AWS IoT Core and how to configure the Shadow Manager component. Next, let's review the MQTT bridge component.

MQTT bridge

The `aws.greengrass.clientdevices.mqtt.Bridge` component is responsible for relaying messages between the three kinds of messaging channels supported by Greengrass. The three kinds of messaging channels are listed here:

- Local IPC topics that components use for messaging
- Local MQTT topics enabled by the Moquette MQTT broker component
- Cloud MQTT topics from using AWS IoT Core

Configuring the MQTT bridge enables you to create flows of messages from a topic on one channel across the boundary to a topic on another channel—for example, copying messages published on the `dt/device1/sensors` IPC topic to the `dt/sensors/all` AWS IoT Core topic for cloud ingestion.

Similar to the Shadow Manager component, each mapping of source and destination topics must be explicitly defined in the component's configuration. To enable the heartbeat messages arriving from the `client.sh` application to the cloud, we used the following configuration in the MQTT bridge:

```
{
  "mqttTopicMapping": {
    "ForwardHeartbeatToCloud": {
      "topic": "dt/localdevice/heartbeat",
      "source": "LocalMqtt",
      "target": "IotCore"
```

```
      }
  }
```

Rather than create one mapping for every single device that used the heartbeat pattern, you could make use of MQTT wildcards and update the topic to dt/+/heartbeat. The best practice is to explicitly define one mapping for each expected device and avoid wildcards unless you have a use case where devices may be migrating between multiple gateway devices and cannot predict specifically which devices may be publishing. Wildcards are great for simplifying manually typed configuration and for legibility but introduce a risk of unanticipated behavior. Wildcards are also not supported for the Pubsub type of topic used for IPC.

Here is another example of using an MQTT bridge to allow delivery of commands from other AWS IoT Core devices or applications directly to a component running on a Greengrass core. The idea of this example is how we might update the settings of all ML-powered components to adjust the rate at which models are used in inference activities. Components subscribed over IPC to the topic can receive updates without requiring a full Greengrass deployment:

```
{
    "mqttTopicMapping": {
      "RemoteUpdateMLComponents": {
          "topic": "cmd/ml-components/inferenceRate",
          "source": "IotCore",
          "target": "Pubsub"
      }
  }
}
```

You have now finished deploying managed components for the purposes of moving messages back and forth between the edge and the cloud. This functionality creates a path for local devices to communicate with each other through the hub device, for components to interact with cloud services, and any other combination of local devices, components, and the cloud exchanging messages. You also learned how AWS IoT Core and the Device Shadow service work to synchronize the state for local devices and components built on the underlying messaging systems. These tools enable you to build edge ML solutions with any kind of messaging requirements without writing any code or managing infrastructure for your messaging needs.

In the next section, you will deploy your first ML model using a combination of prebuilt models, inference code, and a custom component for acting on inference results.

Deploying your first ML model

Now that you are familiar with remote deployments and loading resources from the cloud, it is time to deploy your first ML-powered capability to the edge! After all, a component making use of ML models is much like other components we have deployed. It is a combination of dependencies, runtime code, and static resources that are hosted in the cloud.

Reviewing the ML use case

In this case, the dependencies are packages and libraries for using OpenCV (an open source library for **computer vision (CV)** use cases) and the **Deep Learning Runtime (DLR)**, the runtime code is a preconfigured sample of inference code that uses DLR, and the static resources are a preconfigured model store for image classification and some sample images. The components deployed in this example are all provided and managed by AWS.

The solution that you will deploy simulates the use case for our HBS device hub that performs a simple image classification as part of a home security story for our customers.

Let's say an add-on device that HBS customers can buy for their hub is a security camera that notifies the customer when a person has approached the front door. A camera device takes a picture on an interval or paired with a motion sensor and stores that image as a file on the hub device. The hub device then performs an inference against that image using a local ML model to identify if any person is detected. If so, an audio notification can be played on the hub device or a text notification sent to the customer. Here is an illustration of the solution and its parts:

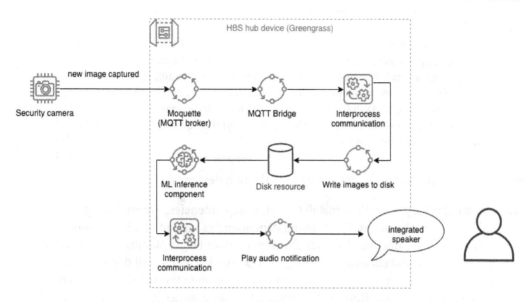

Figure 4.9 – Sample architecture for audio notification of subject recognized by the camera

Today, this kind of security capability has many products in the market. However, what is not yet common is performing the ML inference at the edge. This presents a key opportunity for differentiation because the **total cost of ownership** (**TCO**) goes down by saving bandwidth on image transfer to the cloud and saving the compute costs of running the ML inference in the cloud. The security solution will remain operational even if the customer's internet service is disrupted and has no dependency on the availability of a cloud service. There are also tangible benefits for processing and notification latency, as well as a privacy story that appeals to customers since any image data is kept locally.

This use case demonstrates all four of the key benefits for ML deployed to the edge that we covered in *Chapter 1, Introduction to the Data-Driven Edge with Machine Learning*: latency, availability, cost, and governance.

Steps to deploy the ML workload

The steps to take for this section are pretty simple, thanks to the managed components provided by AWS. There are three components to deploy from the AWS catalog and one more custom component to process the inference results. Since the target prototype hardware of Raspberry Pi and Sense HAT does not include any camera interface, we will lean on the sample image provided in the managed component to emulate the sensor. The one custom component serves as our actuator, and it will display the inference results on the **light-emitting diode** (**LED**).

> **Note**
>
> If you are not using the target Raspberry Pi and Sense HAT hardware, you must alter the steps to use the `com.hbs.hub.ScrollMessageSimulated` component instead. Alternatively, you can view inference results using the AWS IoT Core management console (covered in *Step 5*), any other MQTT client connecting to AWS IoT Core (subscribe to the topic marked in *Step 5*), or review the inference component logs (locally on the device or add these logs to the `LogManager` configuration).

The following steps deploy the ML solution to the hub device:

1. Update the Raspberry Pi to install OpenCV dependencies. We are taking a shortcut here by manually installing dependencies instead of adding complexity to the example (more on this after the steps). Other target platforms should only need `glibc` installed, and the managed components will install dependencies as needed. Open an `ssh` connection to your Pi or open the Terminal program directly on the device. Run the following command: `sudo apt-get install libopenjp2-7 libilmbase23 libopenexr-dev libavcodec-dev libavformat-dev libswscale-dev libv4l-dev libgtk-3-0 libwebp-dev`.

2. Create a new custom component in your AWS account from the book repository's source for the `com.hbs.hub.ScrollMessage` component. These are the same *1-4 steps* from the *Registering a component in IoT Greengrass* section. Upload the `archive.zip` file for this component to S3, edit the recipe file to point at your artifact in S3, then register the new custom component.

3. Edit the `deployment-ml.json` file to update the `targetArn` property, as in past sections. Save the file.

4. Create a new deployment using the `deployment-ml.json` file, as follows: `aws greengrassv2 create-deployment --cli-input-json file:// deployment-ml.json`.

5. When the deployment concludes (it can possibly take a few hours to install the bundled DLR dependencies on the Pi), you should start seeing new inference results appear on your Sense HAT LED matrix. For readers not using the Raspberry Pi with Sense HAT, you can view results using the AWS IoT Core management console. Log in to `https://console.aws.amazon.com/iot`, navigate to **Test** > **MQTT test client**, and subscribe to the `ml/dlr/image-classification` ML inference topic.

If you are using the AWS IoT Core console or an MQTT client to check the inference results, they should look something like this:

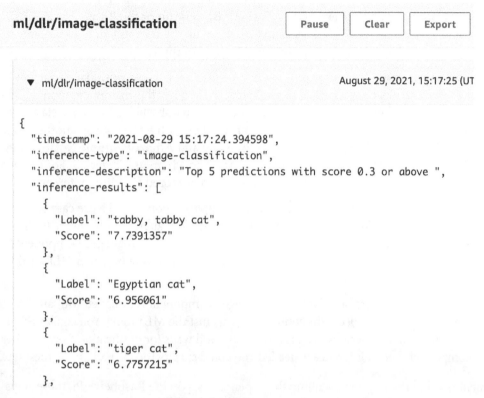

```
ml/dlr/image-classification          Pause    Clear    Export


   ▼  ml/dlr/image-classification                August 29, 2021, 15:17:25 (UT

{
  "timestamp": "2021-08-29 15:17:24.394598",
  "inference-type": "image-classification",
  "inference-description": "Top 5 predictions with score 0.3 or above ",
  "inference-results": [
    {
      "Label": "tabby, tabby cat",
      "Score": "7.7391357"
    },
    {
      "Label": "Egyptian cat",
      "Score": "6.956061"
    },
    {
      "Label": "tiger cat",
      "Score": "6.7757215"
    },
```

Figure 4.10 – Screenshot of the inference results in AWS IoT Core

You have now deployed your first ML workload to the edge! This workload will continue to run at a 10-second interval and will process the file found at /greengrass/ v2/packages/artifacts-unarchived/component-name/image_ classification/sample_images/cat.jpeg on the hub device. There are a number of ways to extend this example for your own testing and use cases, such as the following:

- **Pass in alternate images to the inference component**: You can merge in a new configuration for the aws.greengrass.DLRImageClassification component to specify where to source the processed image—for example, you can specify a new directory with the ImageDirectory parameter and a filename with the ImageName parameter. Then, you could create a custom component that includes your own images as artifacts and rotates them through that filename at that path to see what the classification engine identifies.

- **Hook up a camera as input**: You can merge in a new configuration for the `aws.greengrass.DLRImageClassification` component to specify a camera as the source for images to classify. For prototyping on the Raspberry Pi device, you can connect a **Universal Serial Bus** (**USB**) webcam or the **Raspberry Pi Camera Module v2**. More details on how to set up the component to use a camera can be found in the *References* section at the end of this chapter.

- **Escalate an event for a type of classified result**: The current workload reports all inference results on a specified topic. You could modify the custom component that subscribes to these results and only update the actuator when a human is classified in the inference results. Alternatively, you could publish a message on another topic reserved for alerts and configure AWS IoT Core and **Amazon Simple Notification Service** (**Amazon SNS**) to send you an alert when a human is detected. For these events, you may also want the detected image to be stored for future reference.

- **Deploy object detection workload**: This example demonstrated a use case for classifying the primary object in the source image. An alternate use case is to list all objects detected in the source image. AWS also provides managed components for object detection workloads and for CV use cases running on either DLR or the TensorFlow framework.

- **Notify of new images available**: The managed component is designed with an interval pattern to process the source image against the ML model. You could design a variation that subscribes to a local IPC topic and waits for notification from another component that an inference is needed and could specify the file path in the message.

As mentioned in the step for installing the dependencies on the Raspberry Pi, these steps take a shortcut that doesn't reflect the best practices of our solution design. Rather than manually installing dependencies on the device, a better way is to use IoT Greengrass to install those missing dependencies. This section could have walked you through forking the AWS managed components in order to add more dependencies in the install lifecycle step, but for the purposes of this exercise was deemed not a valuable detour.

Furthermore, the default behavior of the `aws.greengrass.DLRImageClassification` managed component publishes the inference results to a cloud topic on AWS IoT Core instead of a local topic on IPC. For workloads that report all activity to the cloud, this is the desired behavior. However, given the best practices outlined in this book to perform data analysis locally and decouple our components, it would be preferred to publish on a local IPC topic and let another component decide which messages to escalate to the cloud.

This could not be achieved with a simple configuration update as defined in version 2.1.3 of the component. Again, a detour to fork the component just to swap out the cloud publish for a local publish was not deemed valuable for the purposes of this exercise, so the cloud publish behavior is left as-is.

With your first ML-powered workload deployed, you now have an understanding of how ML components are deployed to the edge as a combination of dependencies, code, and static resources.

In this section, you deployed three components as a decoupled set of dependencies, runtimes, and models, plus one more custom component that acts upon the inference results of your ML model. Together, you have all the basics for extending these components to deliver your own image classification and object detection workloads. Next, let's summarize everything you have learned in this chapter.

Summary

This chapter taught you the key differences in remote deploying of components to your Greengrass devices, how to accelerate your solution development using managed components provided by AWS, and getting your first ML workload on your prototype hub device. At this point, you have all the basics you need to start writing your own components and designing edge solutions. You can even extend the managed ML components to get started with some basic CV projects. If you have trained ML models and inference code being used in your business today as containerized code, you could get started with deploying them to the edge now as custom components.

In the next chapter, *Chapter 5*, *Ingesting and Streaming Data from the Edge*, you will learn more about how data moves throughout the edge in prescriptive structures, models, and transformations. Proper handling of data at the edge is important for adding efficiency, resilience, and security to your edge ML solutions.

Knowledge check

Before moving on to the next chapter, test your knowledge by answering these questions. The answers can be found at the end of the book:

1. What are some examples that differentiate static and dynamic resources of an edge component?

2. Where are your components' artifacts stored so that they can be referenced by your recipe files?

3. Can you modify an artifact stored in the cloud after it has been included in a registered custom component?

4. Why can't you write to artifact files that have been loaded onto the edge device through deployment?

5. True or false: Edge devices can belong to multiple thing groups in the cloud AWS IoT Core service and each thing group can have one active Greengrass deployment associated with it.

6. Can you think of a use case for one edge device to receive deployments from multiple thing groups?

7. True or false: A single deployment can reset a component's configuration and apply a merge of a new configuration.

8. Which of the following managed components is responsible for deploying a local MQTT broker and connecting leaf devices?

    MQTT bridge/Moquette/device shadows

9. Which of the following managed components is responsible for synchronizing the state between the edge and the cloud?

    MQTT bridge/Moquette/device shadows

10. Which of the following managed components is responsible for relaying messages between communications channels such as MQTT, IPC, and the cloud?

    MQTT bridge/Moquette/device shadows

11. How might you design the flow of data from a local IoT sensor to an application running ML inference and play an alarm sound over an attached loudspeaker? Think about the different communications channels available and our best practice of decoupled services. Try sketching it out on paper as if you were proposing the design to a colleague.

References

Take a look at the following resources for additional information on the concepts discussed in this chapter:

- *Perform sample image classification inference on images from a camera using TensorFlow Lite*: https://docs.aws.amazon.com/greengrass/v2/developerguide/ml-tutorial-image-classification-camera.html

- *AWS-provided components*: https://docs.aws.amazon.com/greengrass/v2/developerguide/public-components.html

- OpenCV website: https://opencv.org/

- *AWS IoT Device Shadow service*: https://docs.aws.amazon.com/iot/latest/developerguide/iot-device-shadows.html

- *DLR: Compact Runtime for Machine Learning Models*: https://neo-ai-dlr.readthedocs.io

- TensorFlow website: https://www.tensorflow.org/

5
Ingesting and Streaming Data from the Edge

Edge computing can reduce the amount of data transferred to the cloud (or on-premises datacenter), thus saving on network bandwidth costs. Often, high-performance edge applications require local compute, storage, network, data analytics, and machine learning capabilities to process high-fidelity data in low latencies. AWS extends infrastructure to the edge, beyond **Regions** and **Availability Zones**, as close to the endpoint as required by your workload. As you will have learned in previous chapters, **AWS IoT Greengrass** allows you to run sophisticated edge applications on devices and gateways.

In this chapter, you will learn about the different data design and transformation strategies applicable for edge workloads. We will explain how you can ingest data from different sensors through different workflows based on **data velocity** (such as hot, warm, and cold), **data variety** (such as structured and unstructured), and **data volume** (such as high frequency or low frequency) on the edge. Thereafter, you will learn the approaches of streaming the raw and transformed data from the edge to different cloud services. By the end of this chapter, you should be familiar with data processing using AWS IoT Greengrass.

In this chapter, we're going to cover the following main topics:

- Defining data models for IoT workloads

- Designing data patterns for the edge

- Getting to know Stream Manager

- Building your first data orchestration workflow on the edge

- Streaming from the edge to a data lake on the cloud

Technical requirements

The technical requirements for this chapter are the same as those outlined in *Chapter 2, Foundations of Edge Workloads*. See the full requirements in that chapter.

You will find the GitHub code repository here: `https://github.com/ PacktPublishing/Intelligent-Workloads-at-the-Edge/tree/main/ chapter5`

Defining data models for IoT workloads

According to the IDC, the sum of the world's data will grow from 33 **zettabytes** (**ZB**) in 2018 to 175 ZB by 2025. Additionally, the IDC estimates that there will be 41.6 billion connected IoT devices or *things*, generating 79.4 ZB of data in 2025 (`https://www.datanami. com/2018/11/27/global-datasphere-to-hit-175-zettabytes-by-2025- idc-says/`). Additionally, many other sources reiterate that data and information are the *currency*, the *lifeblood*, and even the *new oil* of the information industry.

Therefore, the data-driven economy is here to stay and the **Internet of Things** (**IoT**) will act as the enabler to ingest data from a huge number of devices (or endpoints), such as sensors and actuators, and generate aggregated insights for achieving business outcomes. Thus, as an IoT practitioner, you should be comfortable with the basic concepts of **data modeling** and how that enables **data management** on the edge.

All organizations across different verticals such as industrial, commercial, consumer, transportation, energy, healthcare, and others are exploring new use cases to improve their top line or bottom line and innovate on behalf of their customers. IoT devices such as a connected hub in a consumer home, a smart parking meter on a road, or a connected car will coexist with customers and will operate even when there is no connectivity to the internet.

This is a paradigm shift from the centralized solutions that worked for enterprises in the past. For example, a banking employee might have hosted their workloads in datacenters, but now they can monitor customer activities (such as suspicious actions, footfalls, or availability of cash in an ATM) at their branch locations in near real time to serve customers better. Therefore, a new strategy is required to act on the data generated locally and be able to process and stream data from the edge to the cloud.

In this chapter, we are going to rethink and re-evaluate the applicability of different big data architectures in the context of IoT and edge computing. The three areas we will consider are data management, data architecture patterns, and anti-patterns.

What is data management?

As per the **Data Management Body of Knowledge (DMBOK2)** from the **Data Management Association (DAMA)**, data management is the development, execution, and supervision of plans, policies, programs, and practices that deliver, control, protect, and enhance the value of data and information assets throughout their life cycles (for more information, please refer to *DAMA-DMBOK2* at https://technicspub.com/dmbok/).

DAMA covers the data management framework in great detail, as shown in the following diagram:

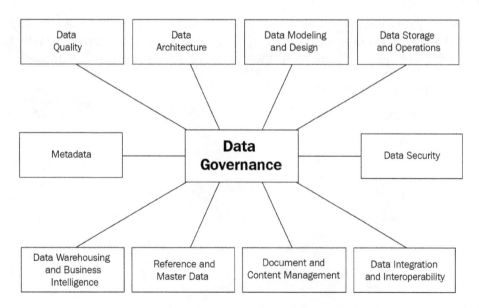

Figure 5.1 – The data management life cycle

Here, we recognized an opportunity to augment the framework from DAMA with concepts that are relevant to edge computing. Therefore, in this section, we will dive deeper into the principles related to data modeling, data architecture, and **Data Integration and Interoperability (DII)**, which we think are relevant for edge computing and IoT.

Let's define data in the context of IoT before we discuss how to model it. IoT data is generated from different sensors, actuators, and gateways. Therefore, they can come in different forms such as the following:

- **Structured data**: This refers to a predictable form of data; examples include device metadata and device relationships.

- **Semi-structured data**: This is a form of data with a certain degree of variance and randomness; examples include sensor and actuator feeds.

- **Unstructured data**: This is a form of data with a higher degree of variance and randomness; examples include raw images, audio, or videos.

Now, let's discuss how the different forms of data can be governed, organized, and stored using data modeling techniques.

What is data modeling?

Data modeling is a common practice in software engineering, where data requirements are defined and analyzed to support the business processes of different information systems in the scope of the organization. There are three different types of data models:

- Conceptual data models
- Logical data models
- Physical data models

In the following diagram, the relationships between different modeling approaches are presented:

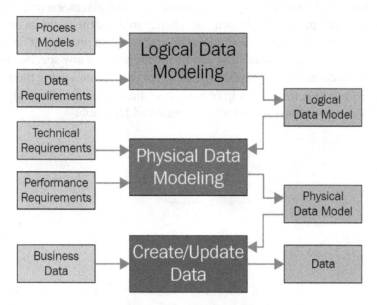

Figure 5.2 – Data modeling approaches

Fun fact

The Enigma machine was used by the German military as the primary mode of communication for all secure wireless communications during World War II. Alan Turing cracked the Enigma code roughly 80 years ago when he figured out the text that's placed at the end of every message. This helped to decipher key secret messages from the German military and helped end the world war. Additionally, this mechanism led to the era of unlocking insights by defining a language to decipher data, which was later formalized as data modeling.

The most common data modeling technique for a database is an **Entity-Relationship** **(ER) model**. An ER model in software engineering is a common way to represent the relationship between different entities such as people, objects, places, events, and more in a graphical way in order to organize information better to drive the business processes of an organization. In other terms, an ER model is an abstract data model that defines a data structure for the required information and can be independent of any specific database technology. In this section, we will explain the different data models using the ER model approach. First, let's define, with the help of a use case diagram, the relationship between a customer and their devices associated with a connected HBS hub:

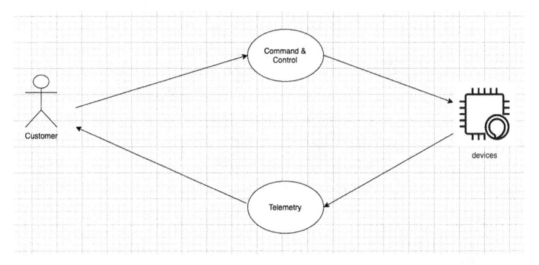

Figure 5.3 – A use case diagram

Now, let's build the ER diagram through a series of conceptual, logical, and physical models:

1. The **conceptual data model** defines the entities and their relationships. This is the first step of the data modeling exercise and is used by personas such as data architects to gather the initial set of requirements from the business stakeholders. For example, *sensor*, *device*, and *customer* are three entities in a relationship, as shown in the following diagram:

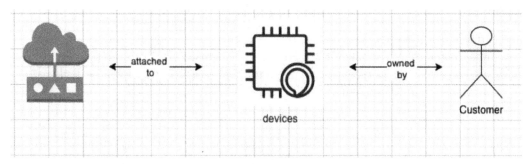

Figure 5.4 – A conceptual data model

2. The conceptual model is then rendered into a **logical data model**. In this step of data modeling, the data structure along with additional properties are defined using a conceptual model as the foundation.

    For example, you can define the properties of the different entities in the relationship such as the sensor type or device identifier (generally, a serial number or a MAC address), as shown in the following list. Additionally, there could be different forms of relationships, such as the following:

    ▪ Association is the relationship between devices and sensors.

    ▪ Ownership is the relationship between customers and devices.

    The preceding points are illustrated in the following diagram:

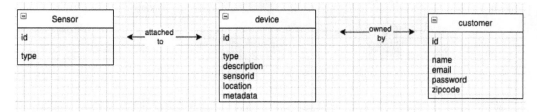

Figure 5.5 – A logical data model

3. The final step in data modeling is to build a **physical data model** from the defined logical model. A physical model is often a technology or product-specific implementation of the data model. For example, you define the data types for the different properties of an entity, such as a number or a string, that will be deployed on a database solution from a specific vendor:

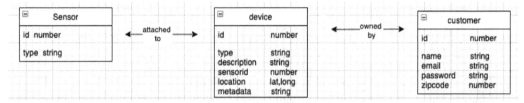

Figure 5.6 – A physical data model

Enterprises have used ER modeling for decades to design and govern complex distributed data solutions. All the preceding steps can be visualized as the following workflow, which is not limited to any specific technology, product, subject area, or operating environment (such as a data center, cloud, or edge):

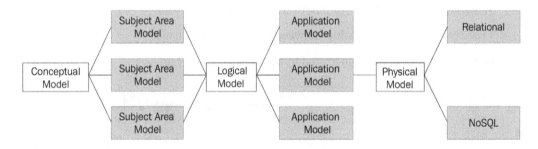

Figure 5.7 – The data modeling flow

Now that you have understood the foundations of data modeling, in the next section, let's examine how this can be achieved for IoT workloads.

How do you design data models for IoT?

Now, let's take a look at some examples of how to apply the preceding data modeling concepts to the realm of structured, unstructured, and semi-structured data that are common with IoT workloads. Generally, when we refer to data modeling for structured data, a **relational database** (**RDBMS**) comes to mind first. However, for most IoT workloads, structured data generally includes hierarchical relationships between a device and other entities. And that is better illustrated using a graph or an ordered key-value database. Similarly, for semi-structured data, when it comes to IoT workloads, it's mostly illustrated as a key-value, time series, or document store.

In this section, we will give you a glimpse of data modeling techniques using NoSQL data solutions to continue building additional functionalities for HBS. Modeling an RDBMS is outside the scope of this book. However, if you are interested in learning about them, there are tons of materials available on the internet that you can refer to.

NoSQL databases are designed to offer freedom to developers to break away from a longer cycle of database schema designs. However, it's a mistake to assume that NoSQL databases lack any sort of data model. Designing a NoSQL solution is quite different from an RDBMS design. For RDBMS, developers generally create a data model that adheres to normalization guidelines, without focusing on access patterns. The data model can be modified later when new business requirements arise, thus leading to a lengthy release cycle. The collected data is organized in different tables with rows, columns, and referential integrities. In contrast, for a NoSQL solution design, developers cannot begin designing the models until they know the questions that are required to be answered. Understanding the business queries working backward from the use case is quintessential. Therefore, the general rule of thumb to remember during data modeling through a relational or NoSQL database is the following:

- Relational modeling primarily cares about the structure of data. The design principle is *What answers do I get?*

- NoSQL modeling primarily cares about application-specific access patterns. The design principle is *What questions do I ask?*

> **Fun fact**
>
> The common translation of the NoSQL acronym is *Not only SQL*. This highlights the fact that NoSQL doesn't only support NoSQL, but it can handle relational, semi-structured, and unstructured data. Organizations such as Amazon, Facebook, Twitter, LinkedIn, and Google have designed different NoSQL technologies.

Before I show you some examples of data modeling, let's understand the five fundamental properties of our application's (that is, the HBS hub) access patterns that need to be considered in order to come up with relevant questions:

- **Data type**: What's the type of data in scope? For example, is the data related to telemetry, command-control, or critical events? Let's quickly refresh the use of each of these data types:

  a) **Telemetry**: This is a constant stream of data transmitted by sensors/actuators, such as temperature or humidity readings , which can be aggregated on the edge or published as it is to the cloud for further processing.

b) **Command and Control**: These are actionable messages, such as turning on/off the lights, which can occur between two devices or between an end user and the device.

c) **Events**: These are data patterns that identify more complex scenarios than regular telemetry data, such as network outages in a home or a fire alarm in a building.

- **Data size**: What is the quantity of data in scope? Is it necessary to store and retrieve data locally (on the edge), or does the data require transmission to a different data persistence layer (such as a data lake on the cloud)?

- **Data shape**: What's the form of data being generated from different edge devices such as text, blobs, and images? Note that different data forms such as images and videos might have different computational needs (think of GPUs).

- **Data velocity**: What's the speed of data to process queries based on the required latencies? Do you have a hot, warm, or cold path of data?

- **Data consistency**: How much of this data needs to have strong versus eventual consistency?

Answering the preceding questions will help you to determine whether the solution should be based on one of the following:

- **BASE methodology**: **Basically Available**, **Soft-state**, **Eventual consistency**, which are typical characteristics of NoSQL databases

- **ACID methodology**: **Atomicity**, **Consistency**, **Isolation**, and **Durability**, which are typical characteristics of relational databases

We will discuss these concepts in more detail next.

Selecting between ACID or BASE for IoT workloads

The following table lists some of the key differences between the two methodologies. This should enable you to make an informed decision working backward from your use case:

Functionality	ACID	BASE
Outcome	Schema driven	Access pattern driven
Data Structure	Schema must exist	Dynamic
	Table Structure exists	Created on the fly
	Row / Column oriented	Key value, Column, Document oriented
Consistency	Strong	Eventual, Strong or None
Scale	Product dependent	Highly scalable

Figure 5.8 – ACID versus BASE summary

Fun fact

ACID and BASE represent opposing sides of the pH spectrum. Jim Grey conceived the idea in 1970 and subsequently published a paper, called *The Transaction Concept: Virtues and Limitations*, in June 1981.

So far, you have understood the fundamentals of data modeling and design approaches. You must be curious about how to relate those concepts to the connected HBS product, which you have been developing in earlier chapters. Let's explore how the rubber meets the road.

Do you still remember the **first phase** of data modeling?

Bingo! Conceptual it is.

Conceptual modeling of the connected HBS hub

The following diagram is a hypothetical conceptual model of the HBS hub:

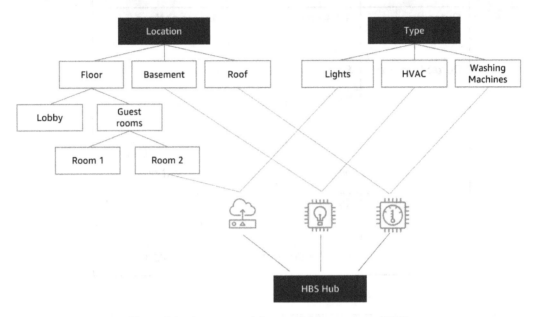

Figure 5.9 – A conceptual data model for connected HBS

In the preceding diagram, you can observe how the different devices such as lights, HVAC, and washing machines are installed in different rooms of a house. Now the conceptual model is in place, let's take a look at the logical view.

The logical modeling of the connected HBS hub

To build the logical model, we need to ask ourselves the type of questions an end consumer might ask, such as the following:

- Show the status of a device (such as is the washing complete?).
- Turn a device on or off (such as turn off the lights).
- Show the readings of a device (such as what's the temperature now?).
- Take a new reading (such as how much energy is being consumed by the refrigerator?).
- Show the aggregated connectivity status of a device (or devices) for a period.

To address these questions, let's determine the access patterns for our end application:

Question	Data Type	Data Size	Data Shape	Data velocity	Data consistency	Data access
Capture the status of the device.	Telemetry	1 KB	Structured	Hot	Eventual	Read
Show me the readings.	Telemetry	5 KB	Semi-structured	Hot	Eventual	Read
Show the aggregated connectivity status.	Telemetry	1 KB	Structured	Warm	Eventual	Read
Turn a device on or off.	Command/Control	1 KB	Structured	Hot	Eventual	Write
Request the device to capture a new reading.	Command/Control	10 KB	Semi-structured	Hot	Eventual	Write

Figure 5.10 – A logical data model for connected HBS

Now that we have captured the summary of our data modeling requirements, you can observe that the solution needs to ingest data in both structured and semi-structured formats at high frequency. Additionally, it doesn't require strong consistency. Therefore, it makes sense to design the data layer using a NoSQL solution that leverages the BASE methodology.

The physical modeling of the connected HBS hub

As the final step, we need to define the physical data model from the gathered requirements. To do that, we will define a primary and a secondary key. You are not required to define all of the attributes if they're not known to you, which is a key advantage of a NoSQL solution.

Defining the primary key

As the name suggests, this is one of the required attributes in a table, which is often known as a **Partition key**. In a table, no two primary keys should have the same value. There is also a concept of a **Composite key**. It's composed of two attributes, a partition key and a **Sort key**.

In our scenario, we will create a `Sensor` table with a composite key (as depicted in the following screenshot). The primary key is a **device identifier**, and the sort key is a timestamp that enables us to query data in a time range:

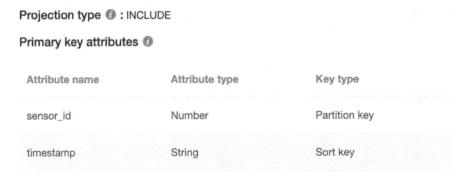

Primary key attributes ⓘ

Attribute name	Attribute type	Key type
device_id	String	Partition key
timestamp	String	Sort key

Figure 5.11 – Composite keys in a sensor table

Defining the secondary indexes

A **secondary index** allows us to query the data in the table using a different key, in addition to queries against the primary or composite keys. This gives your applications more flexibility in querying the data for different use cases. Performing a query using a secondary index is pretty similar to querying from the table directly.

Therefore, for secondary indexes, as shown in the following chart, we have selected the primary key as a sensor identifier (`sensor_id`) along with timestamp as the sort key:

Projection type ⓘ : INCLUDE

Primary key attributes ⓘ

Attribute name	Attribute type	Key type
sensor_id	Number	Partition key
timestamp	String	Sort key

Figure 5.12 – Secondary indexes in a sensor table

Defining the additional attributes

The key advantage of a NoSQL solution is that there is no enforced schema. Therefore, other attributes can be created on the fly as data comes in. That being said, if some of the attributes are already known to the developer, there is no restriction to include those in the data model:

Attribute name	Attribute type
device_name	String
humidity	Number
temperature	Number
duty_cycles	Number
vibration	Number

Figure 5.13 – Other attributes in a sensor table

Now that the data layer exists, let's create the interfaces to access this data.

Defining the interfaces

Now we will create two different facets for the sensor table. A **facet** is a virtual construct that enables different views of the data stored in a table. The facets can be mapped to a functional construct such as a method or an API for performing various **Create, Read, Update, Delete** (**CRUD**) operations on a table:

- putItems: This facet allows write operations and requires the composite keys at the minimum in the payload.

- getItems: This facet allows read operations that can query items with all or selective attributes.

The following screenshot depicts the `getItems` facet definition:

Attribute name	Attribute type	In facets
device_name	String	getItems
humidity	Number	getItems
temperature	Number	getItems
duty_cycles	Number	getItems
vibration	Number	getItems

Figure 5.14 – The getItems facet definition

So, now you have created the data model along with its interfaces. This enables you to understand the data characteristics that are required to develop edge applications.

Designing data patterns on the edge

As data flows securely from different sensors/actuators on the edge to the gateway or cloud over different protocols or channels, it is necessary for it to be safely stored, processed, and cataloged for further consumption. Therefore, any IoT data architecture needs to take into consideration the data models (as explained earlier), data storage, data flow patterns, and anti-patterns, which will be covered in this section. Let's start with data storage.

Data storage

Big data solutions on the cloud are designed to reliably store terabytes, petabytes, or exabytes of data and can scale across multiple geographic locations globally to provide high availability and redundancy for businesses to meet their **Recovery Time Objective (RTO)** and **Recovery Point Objective (RPO)**. However, edge solutions, such as our very own connected HBS hub solution, are resource-constrained in terms of compute, storage, and network. Therefore, we need to design the edge solution to cater to different time-sensitive, low-latency use cases and hand off the heavy lifting to the cloud. A **data lake** is a well-known pattern on the cloud today, which allows a centralized repository to store data as it arrives, without having to first structure the data. Thereafter, different types of analytics, machine learning, and visualizations can be performed on that data for consumers to achieve better business outcomes. So, what is the equivalent of a data lake for the edge?

Let's introduce a new pattern, called a *data pond*, for the authoritative source of data (that is, the golden source) that is generated and temporarily persisted on the edge. Certain characteristics of a data pond are listed as follows:

- A data pond enables the quick ingestion and consumption of data in a fast and flexible fashion. A data producer is only required to know where to push the data, that is, the local storage, local stream, or cloud. The choice of the storage layer, schema, ingestion frequency, and quality of the data is left to the data producer.

- A data pond should work with low-cost storage. Generally, IoT devices are low in storage; therefore, only highly valuable data that's relevant for the edge operations can be persisted locally. The rest of the data is pushed to the cloud for additional processing or thrown away (if noisy).

- A data pond supports schema on read. There can be multiple streams supporting multiple schemas in a data pond.

- A data pond should support the data protection mechanisms at rest and in encryption. It's also useful to implement role-based access that allows auditing the data trail as it flows from the edge to the cloud.

The following diagram shows an edge architecture of how data collected from different sensors/actuators can be persisted and securely governed in a data pond:

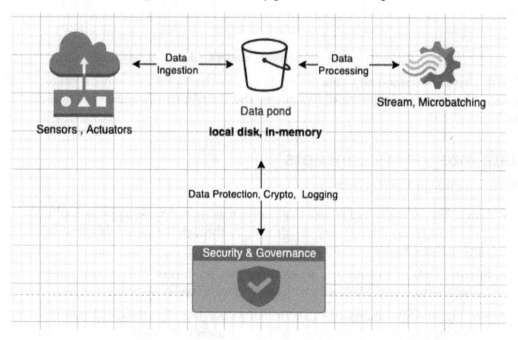

Figure 5.15 – A data pond architecture at the edge

The organizational entities involved in the preceding data flow include the following:

- **Data producers**: These are entities that generate data. These include physical (such as sensors, actuators, or associated devices) or logical (such as applications) entities and are configured to store data in the data pond or publish data to the cloud.

- **Data pond team**: Generally, the data operations team defines the data access mechanisms for the data pond (or lake) and the development team supports data management.

- **Data consumers**: Edge and cloud applications retrieve data from the data pond (or lake) using the mechanisms authorized to further iterate on the data and meet business needs.

The following screenshot shows the organizational entities for the data pond:

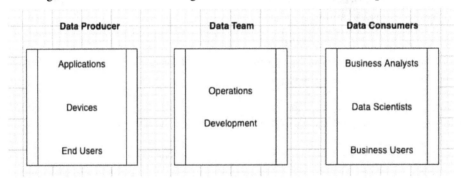

Figure 5.16 – The organizational entities for the data pond

Now you have understood how data can be stored on a data pond and be managed or governed by different entities. Next, let's discuss the different flavors of data and how they can be integrated.

Data integration concepts

DII occurs through different layers, as follows:

- **Batch**: This layer aggregates data that has been generated by data producers. The goal is to increase the accuracy of data insights through the consolidation of data from multiple sources or dimensions.

- **Speed**: This layer streams data generated by data producers. The goal is to allow a near real-time analysis of data with an acceptable level of accuracy.

- **Serving**: This layer merges the data from the batch layer and the speed layer to enable the downstream consumers or business users with holistic and incremental insights into the business.

The following diagram is an illustration of DII:

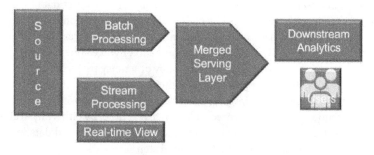

Figure 5.17 -- Data Integration and Interoperability (DII)

As you can see, there are multiple layers within the data flow that are commonly implemented using the **Extract, Transform, and Load** (**ETL**) methodology or the **Extract, Load, and Transform** (**ELT**) methodology in the big data world. The ETL methodology involves steps to extract data from different sources, implement data quality and consistency standards, transform (or aggregate) the data to conform to a standard format, and load (or deliver) data to downstream applications.

The ELT process is a variant of ETL with similar steps. The difference is that extracted data is loaded before the transformation. This is common for edge workloads as well, where the local gateway might not have enough resources to do the transformation locally; therefore, it publishes the data prior to additional processing.

But how are these data integration patterns used in the edge? Let's explore this next.

Data flow patterns

An ETL flow on the edge will include three distinct steps, as follows:

1. Data extraction from devices such as sensors/actuators
2. Data transformation to clean, filter, or restructure data into an optimized format for further consumption
3. Data loading to publish data to the persistence layer such as a data pond, a data lake, or a data warehouse

For an ELT flow, *steps 2* and *3* will take place in the reverse order.

An ETL Scenario for a Connected Home

For example, in a connected home scenario, it's common to extract data from different sensors/actuators, followed by a data transformation that might include format changes, structural changes, semantic conversions, or deduplication. Additionally, data transformation allows you to filter out any noisy data from the home (think of a crying baby or a noisy pet), resulting in reduced network charges of publishing all the bits and bytes to the cloud. Based on a use case such as an intrusion alert or replenishing a printer toner, data transformation can be performed in batch or real time, by eitherphysically storing the result in a staging area or virtually storing the transferred data in memory until you are ready to move to the load step.

These core patterns (ETL or ELT) have evolved, with time, into different data flow architectures, such as event-driven, batch, lambda, and **complex event processing** (*CEP*). We will explain each of them in the next section.

Event-driven (or streaming)

It's very common for edge applications to generate and process data in smaller sets throughout the day when an event happens. Near real-time data processing has a lower latency and can be both synchronous and asynchronous.

In an asynchronous data flow, the devices (such as sensors) do not wait for the receiving system to acknowledge updates before continuing processing. For example, in a connected home, a motion/occupancy sensor can trigger an intruder notification based on a detected event but continue to monitor without waiting for an acknowledgment.

In comparison, in a real-time synchronous data flow, no time delay or other differences between source and target are acceptable. For example, in a connected home, if there is a fire alarm, it should notify the emergency services in a deterministic way.

With AWS Greengrass, you can design both **synchronous** and **asynchronous** data communications. In addition to this, as we build multi-faceted architectures on the edge, it's quite normal to build multiprocessing or multithreaded polyglot solutions on the edge to support different low-latency use cases:

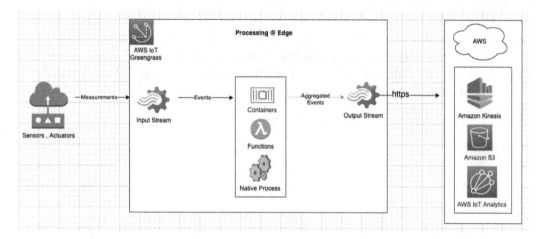

Figure 5.18 – Event-driven architecture at the edge

Micro-batch (or aggregated processing)

Most enterprises perform frequent batch processing to enable end users with business insights. In this mode, data moving will represent either the full set of data at a given point in time, such as the energy meter reading of a connected home at the end of a period (such as the day, week, or month), or data that has changed values since the last time it was sent, such as a hvac reading or a triggered fire alarm. Generally, batch systems are designed to scale and grow proportionally along with the data. However, that's not feasible on the edge due to the lack of horsepower, as explained earlier.

Therefore, for IoT use cases, leveraging micro-batch processing is more common. Here, the data is stored locally and is processed on a much higher frequency, such as every few seconds, minutes, hours, or days (over weeks or months). This allows data consumers to gather insights from local data sources with reduced latency and cost, even when disconnected from the internet. The **Stream Manager** capability of AWS Greengrass allows you to perform aggregated processing on the edge. Stream Manager brings enhanced functionalities regarding how to process data on the edge such as defining a bandwidth and data prioritization for multiple channels, timeout behavior, and direct export mechanisms to different AWS services such as Amazon S3, AWS IoT Analytics, AWS IoT SiteWise, and Amazon Kinesis data streams:

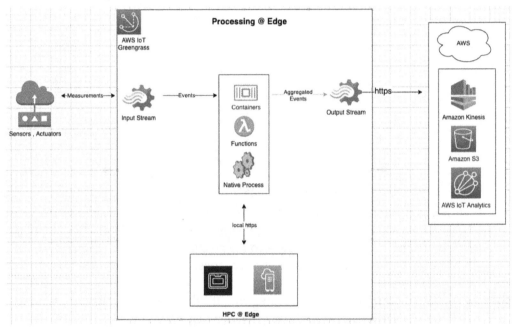

Figure 5.19 – Micro-batch architecture at the edge

Lambda architecture

Lambda architecture is an approach that combines both micro-batch and stream (near real-time) data processing. It makes the consolidated data available for downstream consumption. For example, a refrigeration unit, a humidifier, or any critical piece of machinery on a manufacturing plant can be monitored and fixed before it becomes non-operational. So, for a connected HBS hub solution, micro-batch processing will allow you to detect long-term trends or failure patterns. This capability in turn, will help your fleet operators recommend preventive or predictive maintenance for the machines to end consumers. This workflow is often referred to as the warm or cold path of the data analytics flow.

On the other hand, stream processing will allow the fleet operators to derive near real-time insights through telemetry data. This will enable consumers to take mission-critical actions such as locking the entire house and calling emergency services if any theft is detected. This is also referred to as the hot path in lambda architecture:

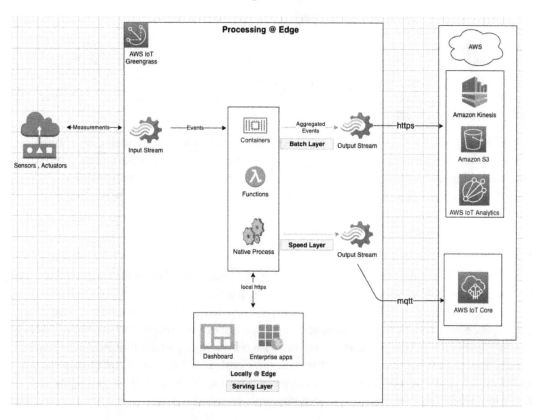

Figure 5.20 – Lambda architecture at the edge

Fun fact

Lambda architecture has nothing to do with the AWS lambda service.
The term was coined by Nathan Marz, who worked on big-data-related technologies at *BackType* and *Twitter*. This is a design pattern for describing data processing that is scalable and fault-tolerant.

Data flow anti-patterns for the edge

So far, we have learned about the common data flow patterns on the edge. Let's also discuss some of the anti-patterns.

Complex Event Processing (CEP)

Events are data patterns that identify complex circumstances from ingested data, such as network outages in a home or a fire alarm in a building. It might be easier to detect events from a few sensors or devices; however, getting visibility into complex events from disparate sources and being able to capture states or trigger conditional logic to identify and resolve issues quickly requires special treatment.

That's where the CEP pattern comes into play. CEP can be resource-intensive and needs to scale to all sizes of data and grow proportionally. Therefore, it's still not a very common pattern on the edge. On the cloud, managed services such as AWS IoT events or AWS EventBridge can make it easier for you to perform CEP on the data generated from your IoT devices.

Batch

Traditionally, in batch processing, data moves in aggregates as blobs or files either on an ad hoc request from a consumer or automatically on a periodic schedule. Data will either be a full set (referred to as snapshot) or a changed set (delta) from a given point in time. Batch processing requires continuous scaling of the underlying infrastructure to facilitate the data growth and processing requirements. Therefore, it's a pattern that is better suited for big data or data warehousing solutions on the cloud. That being said, for an edge use case, you can still leverage the micro-batch pattern (as explained earlier) to aggregate data that's feasible in the context of a resource-constrained environment.

Replication

It's a common practice on the cloud to maintain redundant copies of datasets across different locations to improve business continuity, improve the end user experience, or enhance data resiliency. However, in the context of the edge, **data replication** can be expensive, as you might require redundant deployments. For example, with a connected HBS hub solution, if the gateway needs to support redundant storage for replication, it will increase the **bill of materials** (**BOM**) cost of the hardware, and you can lose the competitive edge on the market.

Archiving

Data that is used infrequently or not actively used can be moved to an alternate data storage solution that is more cost-effective to the organization. Similar to replication, for archiving data locally on the edge, additional deployment of hardware resource is necessary. This increases the **bill of materials** (**BOM**) cost of the device and leads to additional operational overhead. Therefore, it's common to archive the transformed data from the data lake to a cost-effective storage service on the cloud such as **Amazon Glacier**. Thereafter, this data can be used for local operations, data recovery, or regulatory needs.

A hands-on approach with the lab

In this section, you will learn how to build a lambda architecture on the edge using different AWS services. The following diagram shows the lambda architecture:

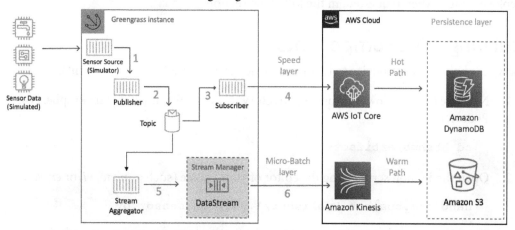

Figure 5.21 – The lab architecture

The preceding workflow uses the following services. In this chapter, you will complete *steps 1–6* (as shown in *Figure 5.21*). This includes designing and deploying the edge components, processing, and transforming data locally, and pushing the data to different cloud services:

Edge Components	Cloud Services
Nucleus	AWS IoT Core
Stream Manager	Amazon S3
Native functions (Python)	

Figure 5.22 – The hands-on lab components

In this hands-on section, your objective will consist of the following:

1. Build the cloud resource (that is, Amazon Kinesis data streams, Amazon S3 bucket, and DynamoDB tables).

2. Build and deploy the edge components (that is, artifacts and recipes) locally on Raspberry Pi.

3. Validate that the data is streamed from the edge to the cloud (AWS IoT Core).

Building cloud resources

Deploy the CloudFormation template from the `chapter5/cfn` folder to create cloud resources such as Amazon S3 buckets, Kinesis data streams, and DynamoDB tables. You will need to substitute these respective names from the *Resources* section of the deployed stack, when requested, in the following section.

Building edge components

Now, let's hop on to our device to build and deploy the required edge components:

1. Navigate to the following working directory from the Terminal of your Raspberry Pi device:

    ```
    cd hbshub/artifacts
    ```

2. Open the Python script using the editor of your choice (such as *nano*, *vi*, or *emac*):

    ```
    nano com.hbs.hub.Publisher/1.0.0/hbs_sensor.py
    ```

    The following code simulates data from the fictional sensors associated with the HBS hub. In the real world, data will be published from the real sensors over serial or GPIO, which needs to be captured. The following function is in a `DummySensor` class that will be referenced by the `Publisher` component in the next step:

    ```python
    def read_value(self):
        message = {}


        device_list = ['hvac', 'refrigerator',
    'washingmachine']
        device_name = random.choice(device_list)


        if device_name == 'hvac' :
            message['device_id'] = "1"
    ```

```
            message['timestamp'] = float("%.4f" %
(time()))
            message['device_name'] = device_name
            message['temperature'] = round(random.
uniform(10.0, 99.0), 2)
            message['humidity'] = round(random.
uniform(10.0, 99.0), 2)
        elif device_name == 'washingmachine' :
            message['device_id'] = "2"
            message['timestamp'] = float("%.4f" %
(time()))
            message['device_name'] = device_name
            message['duty_cycles'] = round(random.
uniform(10.0, 99.0), 2)
        else :
            message['device_id'] = "3"
            message['timestamp'] = float("%.4f" %
(time()))
            message['device_name'] = device_name
            message['vibration'] = round(random.
uniform(100.0, 999.0), 2)


        return message
```

3. Now, open the following `publisher` script and navigate through the code:

```
nano com.hbs.hub.Publisher /1.0.0/hbs_publisher.py
```

The following publisher code streams the data from the dummy sensors, as explained in the previous step, to a `hbs/localtopic` topic every 10 seconds over `ipc`:

```
TIMEOUT = 10
ipc_client = awsiot.greengrasscoreipc.connect()
sensor = DummySensor()


while True:


    message = sensor.read_value()
    message_json = json.dumps(message).encode('utf-8')
```

```
request = PublishToTopicRequest()
request.topic = args.pub_topic
publish_message = PublishMessage()
publish_message.json_message = JsonMessage()
publish_message.json_message.message = message
request.publish_message = publish_message
operation = ipc_client.new_publish_to_topic()
operation.activate(request)
future = operation.get_response()
future.result(TIMEOUT)

print("publish")
time.sleep(5)
```

4. Now you have reviewed the code, check the following recipe file to review the access controls and dependencies that are required by the Publisher component:

```
cd ~/hbshub/recipes
nano com.hbs.hub.Publisher-1.0.0.yaml
```

5. Now we have the component and the recipe, let's create a local deployment:

```
sudo /greengrass/v2/bin/greengrass-cli deployment
create   --recipeDir ~/hbshub/recipes --artifactDir ~/
hbshub/artifacts --merge "com.hbs.hub.Publisher=1.0.0"
Local deployment submitted! Deployment Id: xxxxxxxxxxxxxx
```

6. Verify that the component has successfully been deployed (and is running) using the following command:

```
sudo /greengrass/v2/bin/greengrass-cli component list
```

The following is the output:

```
Components currently running in Greengrass:
Component Name: com.hbs.hub.Publisher
     Version: 1.0.0
     State: RUNNING
```

7. Now that the `Publisher` component is up and running, let's review the code in the `Subscriber` component as well:

```
cd hbshub/artifacts
nano com.hbs.hub.Subscriber/1.0.0/hbs_subscriber.py
```

The `Subscriber` component subscribes to the `hbs/localtopic` topic over the `ipc` protocol and gets triggered by the published events from the publisher:

```
def setup_subscription():
    request = SubscribeToTopicRequest()
    request.topic = args.sub_topic
    handler = StreamHandler()
    operation = ipc_client.new_subscribe_to_
topic(handler)
    future = operation.activate(request)
    future.result(TIMEOUT)
    return operation
```

Then, the subscriber pushes the messages over `mqtt` to the `hbs/cloudtopic` cloud topic on AWS IoT Core:

```
def send_cloud(message_json):
    message_json_string = json.dumps(message_json)
    request = PublishToIoTCoreRequest()
    request.topic_name = args.pub_topic
    request.qos = QOS.AT_LEAST_ONCE
    request.payload = bytes(message_json_string, "utf-8")

    publish_message = PublishMessage()
    publish_message.json_message = JsonMessage()
    publish_message.json_message.message = bytes(message_
json_string, "utf-8")
    request.publish_message = publish_message

    operation = ipc_client.new_publish_to_iot_core()
    operation.activate(request)
    logger.debug(message_json)
```

8. Now you have reviewed the code, let's check the following recipe file to review the access controls and dependencies required by the `Subscriber` component:

```
cd ~/hbshub/recipes
nano com.hbs.hub.Subscriber-1.0.0.yaml
```

9. Now we have the component and the recipe, let's create a local deployment:

```
sudo /greengrass/v2/bin/greengrass-cli deployment
create   --recipeDir ~/hbshub/recipes --artifactDir ~/
hbshub/artifacts --merge "com.hbs.hub.Subscriber=1.0.0"
```

The following is the output:

```
Local deployment submitted! Deployment Id: xxxxxxxxxxxxxx
```

10. Verify that the component has successfully been deployed (and is running) using the following command. Now you should see both the `Publisher` and `Subscriber` components running locally:

```
sudo /greengrass/v2/bin/greengrass-cli component list
```

The following is the output:

```
Components currently running in Greengrass:
Component Name: com.hbs.hub.Publisher
     Version: 1.0.0
     State: RUNNING
Component Name: com.hbs.hub.Subscriber
     Version: 1.0.0
     State: RUNNING
```

11. As you have observed, in the preceding code, the `Subscriber` component will not only subscribe to the local `mqtt` topics on the Raspberry Pi, but it will also start publishing data to AWS IoT Core (on the cloud). Let's verify that from the AWS IoT console:

Please navigate to **AWS IoT console**. | Select **Test** (on the left pane). | Choose **MQTT Client**. | Subscribe to **Topics**. | Type `hbs/cloudtopic`. | Click **Subscribe**.

> **Tip**
> If you have changed the default topic names in the recipe file, please use that name when you subscribe; otherwise, you won't see the incoming messages.

12. Now that you have near real-time data flowing from the edge to the cloud, let's work on the micro-batch flow by integrating with Stream Manager. This component will subscribe to the `hbslocal/topic` topic (same as the subscriber). However, it will append the data to a local data stream using the Stream Manager functionality rather than publishing it to the cloud. Stream Manager is a key functionality for you to build a lambda architecture on the edge. We will break down the code into different snippets for you to understand these concepts better. So, let's navigate to the working directory:

```
nano com.hbs.hub.Aggregator/1.0.0/hbs_aggregator.py
```

13. First, we create a local stream with the required properties such as stream name, data size, time to live, persistence, data flushing, data retention strategy, and more. Data within the stream can stay local for further processing or can be exported to the cloud using the export definition parameter. In our case, we are exporting the data to Kinesis, but you can use a similar approach to export the data to other supported services such as S3, IoT Analytics, and more:

```
iotclient.create_message_stream(
    MessageStreamDefinition(
        name=stream_name,
        max_size=268435456,
        stream_segment_size=16777216,
        time_to_live_millis=None,
        strategy_on_full=StrategyOnFull.OverwriteOldestData,
        persistence=Persistence.File,
        flush_on_write=False,
        export_definition=ExportDefinition(
            kinesis=[
                KinesisConfig(
                    identifier="KinesisExport",
                    kinesis_stream_name=kinesis_stream,
                    batch_size=1,
                    batch_interval_millis=60000,
                    priority=1
```

14. Now the stream is defined, the data is appended through `append_message` api:

```
sequence_number = client.append_message(stream_
name=stream_name, data= event.json_message.message)
```

Fact Check

Stream Manager allows you to deploy a lambda architecture on the edge without having to deploy and manage a separate lightweight database or streaming solution. Therefore, you can reduce the operational overhead or BOM cost of this solution. In addition to this, with Stream Manager as a data pond, you can persist data on the edge using a schema-less approach dynamically (remember BASE?). And finally, you can publish data to the cloud using the native integrations between the Stream Manager and cloud data services, such as IoT Analytics, S3, and Kinesis, without having to write any additional code. Stream Manager can also be beneficial for use cases with larger payloads such as blobs, images, or videos that can be easily transmitted over HTTPS.

---

15. Now that you have reviewed the code, let's add the required permission for the Stream Manager component to update the Kinesis stream:

    Please navigate to **AWS IoT console**. | Select **Secure** (on the left pane). | Choose **Role Aliases** and select the appropriate one. | Click on the **Edit IAM Role**. | Attach policies. | Choose **Amazon Kinesis Full Access**. | Attach policy.

    Please note that it's not recommended to use a blanket policy similar to this for production workloads. This is used here in order to ease the reader into operating in a test environment.

16. Let's perform a quick check of the recipe file prior to deploying this component:

    ```
    cd ~/hbshub/recipes
    nano com.hbs.hub.Aggregator-1.0.0.yaml
    ```

    Please note the `Configuration` section in this recipe file, as it requires the Kinesis stream name to be updated. This can be retrieved from the resources section of the deployed CloudFormation stack. Also, note the dependencies on the Stream Manager component and the reference to `sdk`, which is required by the component at runtime:

    ```
    ComponentConfiguration:
      DefaultConfiguration:
        sub_topic: "hbs/localtopic"
        kinesis_stream: "<replace-with-kinesis-stream-from
    cfn>"
        accessControl:
          aws.greengrass.ipc.pubsub:
            com.hbs.hub.Aggregator:pubsub:1:
    ```

```
            policyDescription: "Allows access to subscribe
to topics"
            operations:
              - aws.greengrass#SubscribeToTopic
              - aws.greengrass#PublishToTopic
            resources:
              - "*"
ComponentDependencies:
  aws.greengrass.StreamManager:
    VersionRequirement: "^2.0.0"
Manifests:
  - Platform:
      os: all
    Lifecycle:
      Install:
        pip3 install awsiotsdk numpy -t .
      Run: |
        export PYTHONPATH=$PYTHONPATH:{artifacts:path}/
stream_manager
        PYTHONPATH=$PWD python3 -u {artifacts:path}/hbs_
aggregator.py --sub-topic="{configuration:/sub_topic}"
--kinesis-stream="{configuration:/kinesis_stream}"
```

17. Next, as we have the artifact and the recipe reviewed, let's create a local deployment:

```
sudo /greengrass/v2/bin/greengrass-cli deployment
create    --recipeDir ~/hbshub/recipes --artifactDir ~/
hbshub/artifacts --merge "com.hbs.hub.Aggregator=1.0.0"
Local deployment submitted! Deployment Id: xxxxxxxxxxxxxx
```

18. Verify that the component has been successfully deployed (and is running) using the following command. You should observe all the following components running locally:

```
sudo /greengrass/v2/bin/greengrass-cli component list
```

The output is as follows:

```
Components currently running in Greengrass:
Component Name: com.hbs.hub.Publisher
    Version: 1.0.0
```

```
     State: RUNNING
  Component Name: com.hbs.hub.Subscriber
     Version: 1.0.0
     State: RUNNING
  Component Name: com.hbs.hub.Aggregator
     Version: 1.0.0
     State: RUNNING
```

19. The `Aggregator` component will publish the data directly from the local stream to the Kinesis stream on the cloud. Let's navigate to the AWS S3 console to check whether the incoming messages are appearing:

    Go to the **Amazon Kinesis console**. | Select **Data Streams**. | Choose the stream. | Go to the **Monitoring** tab. | Check the metrics such as **Incoming data** or **Get records**.

If you see the metrics showing some data points on the chart, it means the data is successfully reaching the cloud.

> **Note**
>
> You can always find the specific resource names required for this lab (such as the preceding Kinesis stream) in the *Resources* or *Output* section of the CloudFormation stack deployed earlier.

Validating the data streamed from the edge to the cloud

In this section, you will perform some final validation to ensure the transactional and batch data streamed from the edge components is successfully persisted on the data lake:

1. In *step 19 of the previous section*, you validated that the Kinesis stream is getting data through metrics. Now, let's understand how that data is persisted to the data lake from the streaming layer:

    Go to the **Amazon Kinesis console**. | Select **Delivery Streams**. | Choose the respective delivery stream. | Click on the **Configuration** tab. | Scroll down to **Destination Settings**. | Click on the S3 bucket under the Amazon S3 destination.

    Click on the bucket and drill down to the child buckets that store the batch data in a zipped format to help optimize storage costs.

2. As the final step, navigate to **AWS DynamoDB** to review the raw sensor data streaming in near real time through AWS IoT Core:

    On the **DynamoDB** console, choose `Tables`. Then, select the table for this lab. Click on **View Items**.

    Can you view the time series data? Excellent work.

> **Note**
>
> If you are not able to complete any of the preceding steps, please refer to the *Troubleshooting* section in the GitHub repository or create an issue for additional instructions.

Congratulations! You have come a long way to learn how to build a lambda architecture that spans from the edge to the cloud using different AWS edge and cloud services. Now, let's wrap up with some additional topics before we conclude this chapter.

Additional topics for reference

Aside from what we have read so far, there are a couple of topics that I wish to mention. Whenever you have the time, please check them out, as they do have lots of benefits and can be found online.

Time series databases

In this chapter, we learned how to leverage a NoSQL (key-value) data store such as Amazon DynamoDB for persisting time series data. Another common way to persist IoT data is to use a **time series database (TSDB)** such as **Amazon Timestream** or **Apache Cassandra**. As you know by now, time series data consists of measurements or events collected from different sources such as sensors and actuators that are indexed over time. Therefore, the fundamentals of modeling a time series database are quite similar to what was explained earlier with NoSQL data solutions. So, the obvious question that remains is *How do you choose between NoSQL and TSDB?* Take a look at the following considerations:

- **Consider the data summarization and data precision requirements**:

    For example, show me the energy utilization on a monthly or yearly basis. This requires going over a series of data points indexed by a time range to calculate a percentile increase of energy over the same period in the last 12 weeks, summarized by weeks. This kind of querying could get expensive with a distributed key-value store.

- **Consider purging the data after a period of time**:

  For example, do consumers really care about the high precision metrics from an hourly basis to calculate their overall energy utilization per month? Probably not. Therefore, it's more efficient to store high-precision data for a short period of time and, thereafter, aggregated and downsampled data for identifying long-term trends. This functionality can partially be achieved with some NoSQL databases as well (such as the DynamoDB item expiry functionality). However, TSDBs are better suited as they can also offer downsampling and aggregation capabilities using different means, such as materialized views.

Unstructured data

You must be curious that most of our discussion in this chapter was related to structured and semi-structured data. We did not touch upon unstructured data (such as images, audio, and videos) at all. You are spot on. Considering IoT is the bridge between the physical world and the cyber world, there will be a huge amount of unstructured data that will need to be processed for different analytics and machine learning use cases.

For example, consider a scenario where the security cameras installed in your customer's home detect any infiltration or unexpected movements through the motion sensors and start streaming a video feed of the surroundings. The feed will be available through your smart hub or mobile devices for consumption. Therefore, in this scenario, the security camera is streaming videos that are unstructured data, as a P2P feed that can also be stored (if the user allows) locally on the hub or to an object store on the cloud. In *Chapter 7, Machine Learning Workloads at the Edge*, you will learn the techniques to ingest, store, and infer unstructured data from the edge. However, we will not delve into the modeling techniques for unstructured data, as it primarily falls under data science and is not very relevant in the day-to-day life of IoT practitioners.

Summary

In this chapter, you learned about different data modeling techniques, data storage, and data integration patterns that are common with IoT edge workloads. You learned how to build, test, and deploy edge components on Greengrass. Additionally, you implemented a lambda architecture to collect, process, and stream data from disparate sources on the edge. Finally, you validated the workflow by visualizing the incoming data on IoT Core.

In the next chapter, you will learn how all this data can be served on the cloud to generate valuable insights for different end consumers.

Knowledge check

Before moving on to the next chapter, test your knowledge by answering these questions. The answers can be found at the end of the book:

1. True or false: Data modeling is only applicable for relational databases.

2. What is the benefit of performing a data modeling exercise?

3. Is there any relevance of ETL architectures for edge computing? (Hint: Think lambda.)

4. True or false: Lambda architecture is the same as AWS lambda service.

5. Can you think of at least one benefit of data processing at the edge?

6. Which component of Greengrass is required to be run at the bare minimum for the device to be functional?

7. True or false: Managing streams for real-time processing is a cloud-only thing.

8. What strategy could you implement to persist data on the edge locally for a longer time?

References

Take a look at the following resources for additional information on the concepts discussed in this chapter:

- *Data Management Body of Knowledge*: https://www.dama.org/cpages/body-of-knowledge

- *Amazon's Dynamo*: https://www.allthingsdistributed.com/2007/10/amazons_dynamo.html

- *NoSQL Design for DynamoDB*: https://docs.aws.amazon.com/amazondynamodb/latest/developerguide/bp-general-nosql-design.html

- *Lambda Architecture*: http://lambda-architecture.net/

- *Managing data streams on the AWS IoT Greengrass Core*: https://docs.aws.amazon.com/greengrass/v2/developerguide/manage-data-streams.html

- *Data Lake on AWS*: https://aws.amazon.com/solutions/implementations/data-lake-solution/

6
Processing and Consuming Data on the Cloud

The value proposition of edge computing is to process data closer to the source and deliver intelligent near real-time responsiveness for different kinds of applications across different use cases. Additionally, edge computing reduces the amount of data that is required to be transferred to the cloud, thus saving on network bandwidth costs. Often, high-performance edge applications require local compute, local storage, network, data analytics, and machine learning capabilities to process high-fidelity data in low latencies. Although AWS IoT Greengrass allows you to run sophisticated edge applications on devices and gateways, it will be resource-constrained compared to the horsepower from the cloud. Therefore, for different use cases, it's quite common to leverage the scale of cloud computing for high-volume complex data processing needs.

In the previous chapter, you learned about the different design patterns around data transformation strategies on the edge. This chapter will focus on explaining how you can build different data workflows on the cloud based on the data velocity, data variety, and data volume collected from the HBS hub running a Greengrass instance. Specifically, you will learn how to persist data in a transactional data store, develop API driven access, and build a serverless data warehouse to serve data to end users. Therefore, the chapter is divided into the following topics:

- Defining big data for IoT workloads
- Introduction to **Domain-Driven Design** (**DDD**) concepts
- Design data flow patterns on the cloud
- Remembering data flow anti-patterns for edge workloads

Technical requirements

The technical requirements for this chapter are the same as those outlined in *Chapter 2, Foundations of Edge Workloads*. See the full requirements in that chapter.

Defining big data for IoT workloads

The term *Big* in **Big data** is relative, as the influx of data has grown substantially in the last two decades from terabytes to exabytes due to the digital transformation of enterprises and connected ecosystems. The advent of big data technologies has allowed people (*think social media*) and enterprises (*think digital transformation*) to generate, store, and analyze huge amounts of data. To analyze datasets of this volume, sophisticated computing infrastructure is required that can scale elastically based on the amount of input data and required outcome. This characteristic of big data workloads, along with the availability of cloud computing, democratized the adoption of big data technologies by companies of all sizes. Even with the evolution of edge computing, big data processing on the cloud plays a key role in IoT workloads, as data is more valuable when it's adjacent and enriched with other data systems. In this chapter, we will learn how the big data ecosystem allows for advanced processing and analytical capabilities on the huge volume of raw measurements or events collected from the edge to enable the consumption of actionable information by different personas.

The integration of IoT with big data ecosystems has opened up a diverse set of analytical capabilities that allows the generation of additional business insights. These include the following:

- **Descriptive analytics**: This type of analytics helps users answer the question of *what happened and why?* Examples of this include traditional queries and reporting dashboards.

- **Predictive analytics**: This form of analytics helps users predict the probability of a given event in the future based on historical events or detected anomalies. Examples of this include early fraud detection in banking transactions and preventive maintenance for different systems.

- **Prescriptive analytics**: This kind of analytics helps users provide specific (clear) recommendations. They address the question of *what should I do if x happens?* Examples of this include an election campaign to reach out to targeted voters or statistical modeling in wealth management to maximize returns.

The outcome of these processes allows organizations to have increased visibility to new information, emerging trends, or hidden data correlation to improve efficiencies or generate new revenue streams. In this chapter, you will learn about the approaches of both descriptive and predictive analytics on data collected from the edge. In addition to this, you will learn how to implement design patterns such as streaming to a data lake or a transactional data store on the cloud, along with leveraging API driven access, which are considered anti-patterns for the edge. So, let's get started with the design methodologies of big data that are relevant for IoT workloads.

What is big data processing?

Big data processing is generally categorized in terms of the three Vs: the volume of data (for example, a terabyte, petabyte, or more), the variety of data (that is, structured, semi-structured, or unstructured), and the velocity of data (that is, the speed with which it's produced or consumed). However, as more organizations begin to adopt big data technologies, there have been additions to the list of Vs, such as the following:

- **Viscosity**: This emphasizes the ease of usability of data; for example, there could be noisy data collected from the edge that's not easy to parse.

- **Volatility**: This refers to how often data changes occur and, therefore, how long the data is useful; for example, capturing specific events at home can be more useful than every other activity.

- **Veracity**: This refers to how trustworthy the data is, for example, if images captured from outdoor cameras are of poor quality, they cannot be trusted to identify intrusion.

For edge computing and the **Internet of Things (IoT)**, all six Vs are relevant. The following diagram presents a visual summary of the range of data that has become available with the advent of IoT and big data technologies. This requires you to consider different ways in which to organize data at scale based on its respective characteristics:

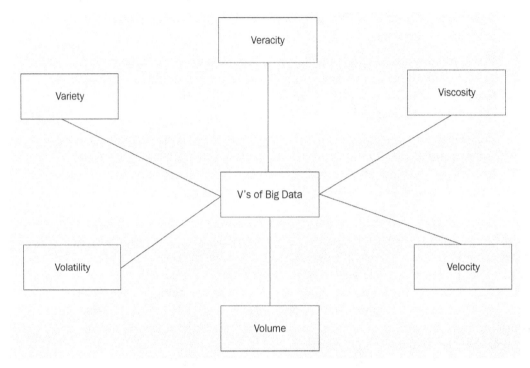

Figure 6.1 – The evolution of big data

So, you have already learned about data modeling concepts in *Chapter 5, Ingesting and Streaming Data from the Edge*, which is a standard way of organizing data into meaningful structures based on data types and relationships and extracting value out of it. However, collecting data, storing it in a stream or a persistent layer, and processing it quickly to take intelligent actions is only one side of the story. The next challenge is to work out how to keep a high quality of data throughout its life cycle so that it continues to generate business value for downstream applications over inconsistency or risk. For IoT workloads, this aspect is critical as the devices or gateways reside in a physical world, with intermittent connectivity, at times, being susceptible to different forms of interference. This is where the domain-driven design (DDD) approach can help.

What is domain-driven design?

To manage the quality of the data better, we need to learn how to organize data by content such as data domains or subject areas. One of the most common approaches in which to do that is through DDD, which was introduced by Eric Evans in 2003. In his book, Eric states *The heart of software is its ability to solve domain-related problems for its user. All other features, vital though they may be, support this basic purpose.* Therefore, DDD is an approach to software development centering around the requirements, rules, and processes of a business domain.

The DDD approach includes two core concepts: bounded context and ubiquitous language. Let's dive deeper into each of them:

- **Bounded context**: Bounded contexts help you to define the logical boundaries of a solution. They can be implemented on the application or business layer, as per the requirements of the organization. However, the core concept is that a bounded context should have its own application, data, and process. This allows the respective teams to clearly define the components they own in a specific domain. These boundaries are important in managing data quality and minimizing data silos, as they grow with different Vs and get redistributed with different consumers within or outside an organization. For example, with a connected HBS solution, there can be different business capabilities required by the internal business functions of HBS and their end consumers. This could include the following:

  - Internal capabilities (for the organizational entities):

    - *Product engineering*: The utilization of different services or features

    - *Fleet operation*: Monitoring fleet health

    - *Information security*: Monitoring the adherence to different regulatory requirements, such as GDPR

    - More such as CRM, ERP, and marketing

  - External capabilities (for the end consumer):

    - *Fleet telemetry*: The processing of data feeds such as a thermostat or HVAC readings from devices in near real time

    - *Fleet monitoring*: Capturing fleet health information or critical events such as the malfunctioning of sensors

    - *Fleet analytics*: Enriching telemetry data with other metadata to perform analysis factoring in different environmental factors such as time, location, and altitude

The following diagram is an illustration of a bounded context:

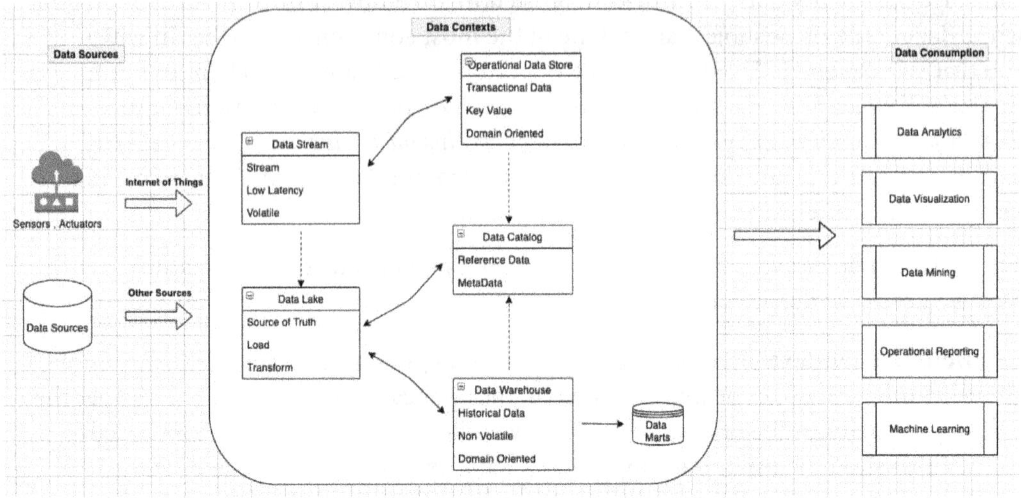

Figure 6.2 – A bounded context

All of these different business capabilities can be defined as a bounded context. So, now the business capabilities have been determined, we can define the technology requirements within this bounded context to deliver the required business outcome. The general rule of thumb is that the applications, data, or processes should be cohesive and not span for consumption by other contexts. In this chapter, we are going to primarily focus on building bounded contexts for external capabilities that are required by end consumers using different technologies.

> **Note**
>
> However, in the real world, there can be many additional factors to bear in mind when it comes to defining a bounded context, such as an organizational structure, product ownership, and more. We will not be diving deep into these factors as they are not relevant to the topic being discussed here.

- **Ubiquitous language**: The second concept in DDD is ubiquitous language. Each bounded context is supposed to have its own ubiquitous language. Applications that belong together within a bounded context should all follow the same language. If the bounded context changes, the ubiquitous language is also expected to be different. This allows the bounded context to be developed and managed by one team and, therefore, aligns with the DevOps methodology as well. This operating model makes it easier for a single team, familiar with the ubiquitous language, to own and resolve different applications or data dependencies quicker.Later in this chapter, you will discover how the different bounded contexts (or workflows) are implemented using a diverse set of languages.

> **Note**
>
> The DDD model doesn't mandate how to determine the bounded context within application or data management. Therefore, it's recommended that you work backward from your use case and determine the appropriate cohesion.

So, with this foundation, let's define some design principles of data management on the cloud – some of these will be used for the remainder of the chapter.

What are the principles to design data workflows using DDD?

We will outline a set of guardrails (that is principles) to understand how to design data workloads using DDD:

- *Principle 1: Manage data ownership through domains* – The quality of the data along with the ease of usability are the advantages of using domains. The team that knows the data best, owns and manages it. Therefore, the data ownership is distributed as opposed to being centralized.

- *Principle 2: Define domains using bounded contexts* – A domain implements a bounded context, which, in turn, is linked to a business capability.

- *Principle 3: Link a bounded context to one or many application workloads* – A bounded context can include one or many applications. If there are multiple applications, all of them are expected to deliver value for the same business capability.

- *Principle 4: Share the ubiquitous language within the bounded context* – Applications that are responsible for distributing data within their bounded context use the same ubiquitous language to ensure that different terminologies and data semantics do not conflict. Each bounded context has a one-to-one relationship with a conceptual data model.

- *Principle 5: Preserve the original sourced data* – Ingested raw data needs to be preserved as a source of truth in a centralized solution. This is often referred to as the golden dataset. This will allow different bounded contexts to repeat the processing of data in the case of failures.

- *Principle 6: Associate data with metadata* – With the growth of data in terms of variety and volume, it's necessary for any dataset to be easily discoverable and classified. This eases the reusability of data by different downstream applications along with establishing a data lineage.

- *Principle 7: Use the right tool for the right job* – Based on the data workflow such as the speed layer or the batch layer, the persistence and compute tools will be different.

- *Principle 8: Tier data storage* – Choose the optimal storage layer for your data based on its access patterns. By distributing the datasets into different storage services, you can build a cost-optimized storage infrastructure.

- *Principle 9: Secure and govern the data pipeline* – Implement control to secure and govern all data at rest and in transit. A mechanism is required to only allow authorized entities to visualize, access, process, and modify data assets. This helps us to protect data confidentiality and data security.

- *Principle 10: Design for Scale* – Last but not least, the cloud is all about the economies of scale. So, take advantage of the managed services to scale elastically and handle any volume of data reliably.

In the remainder of the chapter, we will touch upon most of these design principles (if not all) as we dive deeper into the different design patterns, data flows, and hands-on labs.

Designing data patterns on the cloud

As data flows from the edge to the cloud securely over different channels (such as through speed or batch layers), it is a common practice to store the data in different staging areas or a centralized location based on the data velocity or data variety. These data sources act as a single source of truth and help to ensure the quality of the data for their respective bounded contexts. Therefore, in this section, we will discuss different data storage options, data flow patterns, and anti-patterns on the cloud. Let's begin with data storage.

Data storage

As we learned, in earlier chapters, since edge solutions are constrained in terms of computing resources, it's important to optimize the number of applications or the amount of data persisted locally based on the use case. On the other hand, the cloud doesn't have that constraint, as it comes with virtually unlimited resources with different compute and storage options. This makes it a perfect fit for big data applications to grow and contract based on the demand. In addition to this, it provides easy access to a global infrastructure to orchestrate data required by different downstream or end consumers in the region that is closer to them. Finally, data is more valuable when it's augmented with other data or metadata; thus, in recent times, patterns such as data lakes have become very popular. So, what is a data lake?

A data lake is a centralized, secure, and durable storage platform that allows you to ingest, store structured and unstructured data, and transform the raw data as required. You can think of data lake as a superset of the data pond concepts introduced in *Chapter 5, Ingesting and Streaming Data from the Edge.* Since the IoT devices or gateways are relatively low in storage, only highly valuable data that's relevant for the edge operations can be persisted locally in a data pond:

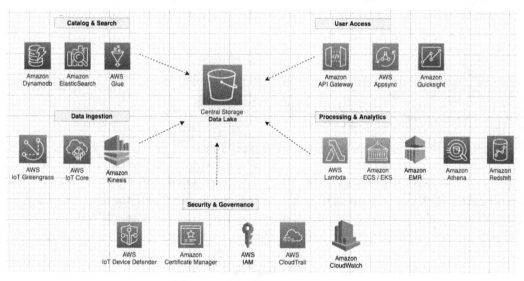

Figure 6.3 – The data lake architecture

Some of the foundational characteristics of the data lake architecture are explained here:

- There is central storage for storing raw data with minimal or no transformation securely. This is a single source of truth of data. The choice of compute, storage layer, schema, ingestion frequency, and data quality is left to the data producer. Amazon S3 is commonly chosen as the central storage since it's a highly scalable, highly durable, and cost-effective service that allows the decoupling of the compute and storage layers. AWS offers different tiering options within Amazon S3 along with a full-fledged archival service referred to as Amazon Glacier.

- There is a persistence layer for storing domain-specific data marts or transformed data in a columnar format (such as Parquet, ORC, or Avro) to achieve isolation by bounded contexts, faster performance, or lower cost. AWS offers different services such as AWS Glue for data transformation and data catalogs, Amazon Athena or Amazon Redshift for data warehouses and data marts, and Amazon EMR or Spark on EMR for managing big data processing.

- There is a persistence layer for storing transactional data ingested from the edge securely. This layer is often referred to as the **Operational Data Store** (**ODS**). AWS offers different services that can be leveraged here based on the given data structures and access patterns, such as Amazon DynamoDB, Amazon RDS, and Amazon Timestream.

You must be wondering how data from a data lake is made available to a data warehouse or an ODS. That's where data integration patterns play a key role.

Data integration patterns

Data Integration and Interoperability (**DII**) happens through the batch, speed, and serving layers. A common methodology in the big data world that intertwines all these layers is **Extract, Transform, and Load** (**ETL**) or **Extract, Load, and Transform** (**ELT**). We have already explained these concepts, in detail, in *Chapter 5, Ingesting and Streaming Data from the Edge*, and discussed how they have evolved with time into different data flow patterns such as event-driven, batch, lambda, and complex event processing. Therefore, we will not be repeating the concepts here. But in the next section, we will explain how they relate to data workflows in the cloud.

Data flow patterns

Earlier in this chapter, we discussed how bounded contexts can be used to segregate different external capabilities for end consumers, such as *fleet telemetry*, *fleet monitoring*, or *fleet analytics*. Now, it's time to learn how these concepts can be implemented using different data flow patterns.

Batch (or aggregated processing)

Let's consider a scenario; you discover that you have been getting a higher electricity bill for the last six months, and you would like to compare the utilization of different equipment for that time period. Alternatively, you want visibility of more granular information, such as how many times did the washing machine run during the day in the last six months? And for how long? This led to how many X watts of consumption?

This is where batch processing helps. It had been the de facto standard of the industry before event-driven architecture gained popularity and is still heavily used for different use cases such as order management, billing, payroll, financial statements, and more. In this mode of processing, a large volume of data, such as thousands or hundreds of thousands of records (or more), is typically transmitted in a file format (such as TXT or CSV), cleaned, transformed, and loaded into a relational database or data warehouse. Thereafter, the data is used for data reconciliation or analytical purposes. A typical batch processing environment also includes a job scheduler that can trigger an analytical workflow based on schedules of feed availability or those that are required by the business.

To design the *fleet analytics* bounded context, we have designed a batch workflow, as follows:

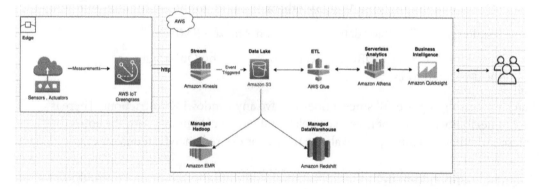

Figure 6.4 – The batch architecture

In this pattern, the following activities are taking place:

- Events streamed from the edge are routed through a streaming service (that is, Amazon Kinesis) to a data lake (that is, Amazon S3).

- Amazon Kinesis allows the preprocessing or enrichment of the data (if required) with additional metadata prior to persisting it to a data lake.

- The data can be crawled or transformed through an ETL engine (that is, AWS Glue) and be easily queried using a serverless analytical service (that is, Amazon Athena). Amazon Athena uses a Presto engine under the hood and is compatible with ANSI SQL.

- Different services such as Amazon S3 and Amazon Athena offer integrations with Amazon QuickSight and different third-party **Business Intelligence** (**BI**) tools through JDBC and ODBC connectors.

- Amazon S3 is highly available and durable object storage that integrates with other big data services such as a fully managed Hadoop cluster (that is, Amazon EMR) or a data warehouse (that is, Amazon Redshift).

> **Fun fact**
> Amazon EMR and Amazon Redshift support big data processing through decoupling of the compute layer and the storage layer, which means there is no need to copy all the data to local storage from the data lake. Therefore, processing becomes more cost-efficient and operationally optimal.

The ubiquitous language used in this bounded context includes the following:

- A REST API for stream processing on Amazon Kinesis, data processing on Amazon S3 buckets, and ETL processing on AWS Glue

- SQL for data analytics on Amazon Athena and Amazon Redshift

- MapReduce or Spark for data processing on Amazon EMR

- Rest APIs, JDBC, or ODBC connectors with Amazon QuickSight or third-party BI tools

Batch processing is powerful since it doesn't have any windowing restrictions. There is a lot of flexibility in terms of how to correlate individual data points with the entire dataset, whether it's terabytes or exabytes in size for desired analytical outcomes.

Event-driven processing

Let's consider the following scenario: you have rushed out of your home, and you get a notification after boarding your commute that you left the cooking stove on. Since you have a connected stove, you can immediately turn it off remotely from an app to avoid fire hazards. Bingo!

This looks easy, but there is a certain level of intelligence required at the local hub (such as HBS hub) and a chain of events to facilitate this workflow. These might include the following:

- Detect from motion sensors, occupancy sensors, or cameras that no one is at home.

- Capture multiple measurements from stove sensors over a period of time.

- Correlate the events to identify this as a hazard scenario using local processes at the edge.

- Stream an event to a message broker and persist it in an ODS.

- Trigger a microservice(s) to notify this event to the end user.

- Remediate the issue based on user response.

So, as you can observe, a lot is happening in a matter of seconds between the edge, the cloud, and the end user to help mitigate the hazard. This is where patterns such as event-driven architectures became very popular in the last decade or so.

Prior to EDA, polling and Webhooks were the common mechanisms in which to communicate events between different components. Polling is inefficient since there is always a lag in terms of how to fetch new updates from the data source and sync them with downstream services. Webhooks are not always the first choice, as they might require custom authorization and authentication configurations. In short, both of these methods require additional work to be integrated or have scaling issues. Therefore, you have the concept of events, which can be filtered, routed, and pushed to different other services or systems with less bandwidth and lower resource utilization since the data is transmitted as a stream of small events or datasets. Similar to the edge, streaming allows the data to be processed as it arrives without incurring any delay.

Generally, event-driven architectures come in two topologies, the mediator topology and the broker topology. We have explained them here:

- **The mediator topology**: There is a need for a central controller or coordinator for event processing. This is generally useful when there is a chain of steps for processing events.

- **The broker topology**: There is no mediator, as the events are broadcast through a broker to different backend consumers.

The broker topology is very common with edge workloads since it decouples the edge from the cloud and allows the overall solution to scale better. Therefore, for the fleet telemetry bounded context, we have designed an event-driven architecture using a broker topology, as shown in the following diagram.

In the following data flow, the events streamed from a connected HBS hub (that is, the edge) are routed over MQTT to an IoT gateway (that is, AWS IoT Core), which allows the filtering of data (if required) through an in-built rules engine and persists the data to an ODS (that is, Amazon DynamoDB). Amazon DynamoDB is a highly performant nonrelational database service that can scale automatically based on the volume of data streamed from millions of edge devices. From the previous chapter, you should already be familiar with how to model data and optimize NoSQL databases for time series data. Once the data is persisted in Amazon DynamoDB, **Create, Read, Update, and Delete (CRUD)** operations can be performed on top of the data using serverless functions (that is, AWS Lambda). Finally, the data is made available through an API access layer (that is, the Amazon API gateway) in a synchronous or asynchronous manner:

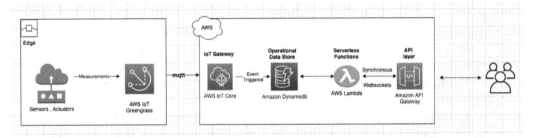

Figure 6.5 – Streaming architecture

The ubiquitous language used in this bounded context includes the following:

- SQL for DynamoDB table access

- Python for developing a lambda function

- A REST API for API gateway and DynamoDB access

Stream processing and EDA is powerful for many IoT use cases that require near-real-time attention such as alerting, anomaly detection, and more, as it analyzes the data as soon as it arrives. However, there is a trade-off with every architecture and EDA is no exception either. With a stream, since the processed results are made available immediately, the analysis of a particular data point cannot consider future values. Even for past values, it's restricted to a shorter time interval, which is generally specified through different windowing mechanisms (such as sliding, tumbling, and more). And that's where batch processing plays a key role.

Complex event processing

Let's consider the following scenario where you plan to reduce food wastage at your home. Therefore, every time you check-in at a grocery store, you receive a notification with a list of perishable items in your refrigerator (or food shelf), as they have not even been opened or are underutilized and are nearing their expiry date.

This might sound like an easy problem to solve, but there is a certain amount of intelligence required at the local hub (such as HBS hub) and a complex event processing workflow on the cloud to facilitate this. It might include the following:

- Based on location sharing and user behavior, the ability to recognize the pattern (or a special event) that the user plans to do grocery shopping.

- Detect from camera sensors installed in the refrigerator or on the food shelves that some of the perishable items are due to expire. Alternatively, use events from smell sensors to detect a pattern of rotten food items.

- Correlate all these patterns (that is, the user, location, and food expiry date) through state machines and apply business rules to identify the list of items that requires attention.

- Trigger a microservice(s) to notify this information to the end user.

This problem might become further complicated for a restaurant business due to the volume of perishable items and the scale at which they operate. In such a scenario, having near real-time visibility to identify waste based on current practices can help the business optimize its supply chain and save a lot of costs. So, as you can imagine, the convergence of edge and IoT with big data processing capabilities such as CEP can help unblock challenging use cases.

Processing and querying events as they arrive in small chunks or in bulk is relatively easier compared to recognizing patterns by correlating events. That's where CEP is useful. It's considered as a subset of stream processing with the focus to identify special (or complex) events by correlating events from multiple sources or by listening to telemetry data for a longer period of time. One of the common patterns to implement CEP is by building state machines.

In the following flow, the events streamed from a connected HBS hub (that is, the edge) are routed over MQTT to an IoT gateway (that is, AWS IoT Core), which filters the complex events based on set criteria and pushes them to different state machines defined within the complex event processing engine (that is, AWS IoT events). AWS IoT events is a fully managed CEP service that allows you to monitor equipment or device fleets for failure or changes in operation and, thereafter, trigger actions based on defined events:

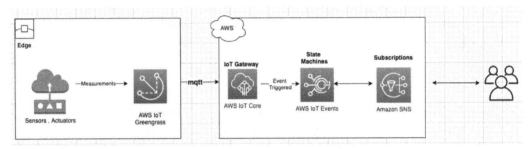

Figure 6.6 – CEP architecture

The ubiquitous language used in the fleet monitoring bounded context includes the following:

- State machines for complex event processing

- A REST API for notifications or subscriptions through **Amazon Simple Notification Service (SNS)**

CEP can be useful for many IoT use cases that require attention based on events from multiple sensors, timelines, or other environmental factors.

There can be many other design patterns that you need to consider in order to design a real-life IoT workload. Those could be functional or non-functional requirements such as data archival for regulatory requirements, data replication for redundancy, and disaster recovery for achieving required RTO or RPO; however, that's beyond the scope of this book, as these are general principles and not necessarily related to edge computing or IoT workloads. There are many other books or resources available on those topics if they are of interest to you.

Data flow anti-patterns for the cloud

The anti-patterns for processing data on the cloud from edge devices can be better explained using three laws – the law of physics, the law of economics, and the law of the land:

- **Law of physics**: For use cases where latency is critical, keeping data processing closer to the event source is usually the best approach, since we cannot beat the speed of light, and thus, the round-trip latency might not be affordable. Let's consider a scenario where an autonomous vehicle needs to apply a hard brake after detecting a pedestrian; it cannot afford the round-trip latency from the cloud. This factor is also relevant for physically remote environments, such as mining, oil, and gas facilities where there is poor or intermittent network coverage. Even for our use case here, with connected HBS, if there is a power or network outage, the hub is still required to be intelligent enough to detect intrusion by analyzing local events.

- **Law of economics**: The cost of compute and storage has reduced exponentially in the last few decades compared to networking cost, which might still become prohibitive at scale. Although digital transformation has led to data proliferation across different industries, much of the data is of low quality. Therefore, local aggregation and the filtering of data on the edge will allow you to publish high-value data to the cloud only reducing networking bandwidth costs.

- **Law of the land**: Most industries need to comply with regulations or compliance requirements related to data sovereignty. Therefore, the local retention of data in a specific facility, region, or country might turn out to be a key factor in the processing of data. Even for our use case here with connected HBS, the workload might need to be in compliance with GDPR requirements.

AWS offers different edge services for supporting use cases that need to comply with the preceding laws and are not limited to IoT services only. For example, consider the following:

- **Infrastructure**: AWS Local Zones, AWS Outposts, and AWS Wavelength

- **Networking**: Amazon CloudFront, and POP locations

- **Storage**: AWS Storage Gateway

- **Rugged and disconnected edge devices**: AWS Snowball Edge and AWS Snowcone

- **Robotics**: AWS Robomaker

- **Video analytics**: Amazon Kinesis Video Streams

- **Machine learning**: Amazon Sagemaker Neo, Amazon Sagemaker Edge Manager, Amazon Monitron, and AWS Panorama

The preceding services are beyond the scope of the book, and the information is only provided for you to be well informed of the breadth and depth of AWS edge services.

A hands-on approach with the lab

In this section, you will learn how to design a piece of architecture on the cloud leveraging the concepts that you have learned in this chapter. Specifically, you will continue to use the lambda architecture pattern introduced in *Chapter 5, Ingesting and Streaming Data from the Edge*, to process the data on the cloud:

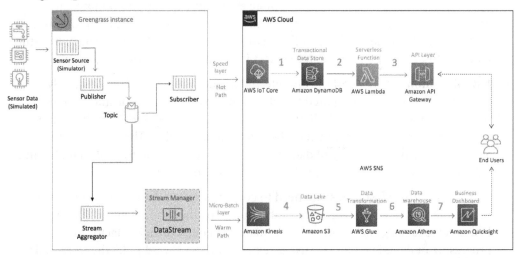

Figure 6.7 – Hands-on architecture

In the previous chapter, you already completed steps 1 and 4. This chapter will help you to complete steps 2, 3, 5, 6, and 7, which includes consuming the telemetry data and building an analytics pipeline for performing BI:

Cloud Services (Bounded Context – fleet telemetry)	Cloud Services (Bounded Context – fleet analytics)
AWS IoT Core	Amazon Kinesis
Amazon DynamoDB	Amazon S3
Lambda functions (Python)	AWS Glue
Amazon API Gateway	Amazon Athena
	Amazon Quicksight

Figure 6.8 – Hands-on lab components

In this hands-on section, your objectives include the following:

1. Query the ODS.
2. Build an API interface layer to enable data consumption.
3. Build an ETL layer for processing telemetry data in the data lake.
4. Visualize the data through a BI tool.

Building cloud resources

This lab builds on top of the cloud resources that you have already deployed in *Chapter 5, Ingesting and Streaming Data from the Edge*. So, please ensure you have completed the hands-on section there prior to proceeding with the following steps here. In addition, please go ahead and deploy the CloudFormation template from the `chapter 6/cfn` folder to create the resources required in this lab, such as the AWS API gateway, lambda functions, and AWS Glue crawler.

> **Note**
>
> Please retrieve the parameters required for this CloudFormation template (such as an S3 bucket) from the *Output* section of the deployed CloudFormation stack of *Chapter 5, Ingesting and Streaming Data from the Edge*.
>
> In addition to this, you can always find the specific resource names required for this lab (such as lambda functions) from the *Resources* or *Output* sections of the deployed CloudFormation stack. It is a good practice to copy those into a notepad locally so that you can refer to them quickly.

Once the CloudFormation has been deployed successfully, please continue with the following steps.

Querying the ODS

Navigate to the AWS console and try to generate insights from the data persisted in the operational (or transactional) data store. As you learned in the previous chapter, all the near real-time data processed is persisted in a DynamoDB table (`packt_sensordata`):

1. To query the data, navigate to **DynamoDB Console**, select **Tables** (from the left-hand pane), click on the table, and then click on **View items**.
2. Click on the **Query** tab. Put a value of 1 into the `device_id` partition key and click on **Run**. This should return a set of data points with all the attributes.

3. Expand the filters section, and add filters to the following attributes:

    ▪ **Attribute name** – `temperature`.

    ▪ **Type** – **Number**.

    ▪ **Condition** – **Greater than or Equal**.

    ▪ **Value** – **50**.

    ▪ Click **Add filter**.

    ▪ **Attribute name** – `humidity`.

    ▪ **Type** – **Number**.

    ▪ **Condition** – **Greater than or Equal**.

    ▪ **Value** – **35**.

    ▪ Click on **Run**.

    Here, the query interface allows you to filter the data quickly based on different criteria. If you are familiar with SQL, you can also try the PartiQL editor, which is on the DynamoDB console.

4. Additionally, DynamoDB allows you to scan an entire table or index, but this is generally an expensive operation, particularly for a large dataset. To scan a table, click on the **Scan** tab (which is adjacent to **Query**) and then click on **Run**.

For better performance and faster response times, we recommend that you use **Query** over **Scan**.

AWS Lambda

In addition to having interactive query capabilities on data, you will often need to build a presentation layer and business logic for various other personas (such as consumers, fleet operators, and more) to access data. You can define the business logic layer using Lambda:

1. Navigate to the **AWS Lambda** console. Click on **Functions** (from the left-hand pane), and choose the function created using the CloudFormation template earlier.

2. Do you remember that we created two facets (get Items and put Items) during the data modeling exercise in *Chapter 5, Ingesting and Streaming Data from the Edge*, to access data? The following is the logic embedded in a lambda function to implement the equivalent functional construct. Please review the code to understand how the get and put functionalities work:

```
try {
    switch (event.routeKey) {
      case "GET /items/{device_id}":
        var nid = String(event.pathParameters.id);
        body = await dynamo
          .query({
            TableName: "<table-name>",
            KeyConditionExpression: "id = :nid",
            ExpressionAttributeValues: {
              ":nid" : nid
            }
          })
          .promise();
        break;
      case "GET /items":
        body = await dynamo.scan({ TableName: "<table-
name>" }).promise();
        break;
      case "PUT /items":
        let requestJSON = JSON.parse(event.body);
        await dynamo
          .put({
            TableName: "<table-name>",
            Item: {
              device_id: requestJSON.id,
              temperature: requestJSON.temperature,
              humidity: requestJSON.humidity,
              device_name: requestJSON.device_name
            }
          })
          .promise();
```

```
        body = `Put item ${requestJSON.id}`;
        break;
      default:
        throw new Error(`Unsupported route: "${event.
routeKey}"`);
    }
  } catch (err) {
    statusCode = 400;
    body = err.message;
  } finally {
    body = JSON.stringify(body);
  }


  return {
    statusCode,
    body,
    headers
  };};
```

Please note that here, we are using lambda functions. This is because serverless functions have become a common pattern to process event-driven data in near real time. Since it alleviates the need for you to manage or operate any servers throughout the life cycle of the application, your only responsibility is to write the code in a supported language and upload it to the Lambda console.

Fun fact

AWS IoT Greengrass provides a Lambda runtime environment for the edge along with different languages such as Python, Java, Node.js, and C++. That means you don't need to manage two different code bases (such as embedded and cloud) or multiple development teams. This will cut down your development time, enable a uniform development stack from the edge to the cloud, and accelerate your time to market.

Amazon API gateway

Now the business logic has been developed using the lambda function, let's create the HTTP interface (aka the presentation layer) using Amazon API gateway. This is a managed service for creating, managing, and deploying APIs at scale:

1. Navigate to the **Amazon API Gateway** console, click on **APIs** (from the left-hand pane), and choose the API (`MyPacktAPI`) created using the CloudFormation template.

2. Expand the **Develop** section (in the left-hand pane). Click on **Routes** to check the created `REST` methods.

3. You should observe the following operations:

   ```
   /items GET - allows accessing all the items on the
   DynamoDB table (can be an expensive operation)
   ```

4. Continue underneath the **Develop** drop-down menu. Click on **Authorization** and check the respective operations. We have not attached any authorizer in this lab, but it's recommended for real-world workloads. API gateway offers different forms of authorizers, including built-in IAM integrations, **JSON Web Tokens** (**JWT**), or custom logic using lambda functions.

5. Next, click on **Integrations** (underneath **Develop**) and explore the different REST operations (such as `/items GET`). On the right-hand pane, you will see the associated lambda functions. For simplicity, we are using the same lambda function here for all operations, but you can choose other functions or targets such as Amazon SQS, Amazon Kinesis, a private resource in your VPC, or any other HTTP URI if required for your real-world use case.

6. There are many additional options offered by API gateway that relate to CORS, reimport/export, and throttling, but they are not considered in the scope of this lab. Instead, we will focus on executing the HTTP APIs and retrieving the sensor data.

7. Click on the **API** tab (in the left-hand pane), copy the invoke URL underneath **Stages**, and run the following commands to retrieve (or `GET`) items from your device Terminal:

   ```
   a. Query all items from the table
   curl https://xxxxxxxx.executeapi.<region>.amazonaws.com/
   items
   ```

You should see a long list of items on our Terminal that has been retrieved from the dynamodb table.

Amazon API gateway allows you to create different types of APIs, and the one configured earlier falls into the HTTP API category that allows access to lambda functions and other HTTP endpoints. Additionally, we could have used the REST APIs here, but the HTTP option was chosen for its simplicity of use, as it can automatically stage and deploy the required APIs without any additional effort and can be more cost-effective. In summary, you have now completed implementing the bounded context of an ODS through querying or API interfaces.

Building the analytics workflow

In this next section, you will build an analytics pipeline on the batch data persisted on Amazon S3 (that is, the data lake). To achieve this, you can use AWS Glue to crawl data and generate a data catalog. Thereafter, you will use Athena for interactive querying and use QuickSight for visualizing data through charts/dashboards.

AWS Glue

AWS Glue is a managed service that offers many ETL functionalities, such as data crawlers, data catalogs, batch jobs, integration with CI/CD pipelines, Jupyter notebook integration, and more. Primarily, you will use the data crawler and cataloging capabilities in this lab. We feel that might be sufficient for IoT professionals since data engineers will be mostly responsible for these activities in the real world. However, if you believe in learning and are curious, please feel free to play with the other features:

1. Navigate to the **AWS Glue** console, click on **Crawlers** (from the left-hand pane), and select the crawler created earlier using the CloudFormation template.

2. Review some of the key attributes of the crawler definition, such as the following:

    -   State: Is the crawler ready to run?

    -   Schedule: Is the frequency of the crawler set correctly?

    -   Data store: S3.

    -   Include path: Is the location of the dataset correct? This should point to the raw sensor data bucket.

    -   Configuration options: Is the table definition being updated in the catalog based on upstream changes?

3. Additionally, Glue allows you to process different data formats through its classifier functionality. You can process the most common data formats such as Grok, XML, JSON, and CSV with its built-in classifiers along with specifying custom patterns if you have data in a proprietary format.

4. Here, the crawler should run on the specified schedule configured through CloudFormation; however, you can also run it on demand by clicking on **Run Crawler**. If you do the same, please wait for the crawler to complete the transition from the **starting** -> **running** -> **stopping** -> **ready** status.

5. Now the data crawling is complete, navigate to **Tables** (from the left-hand pane) and confirm whether a table (or tables) resembling the name of `*packt*` has been created. If you have a lot of tables already created, another quick option is to use the search button and filter on **Database**: `packt_gluedb`.

6. Click on the table to verify the properties, such as the database, the location, the input/output formats, and the table schema. Confirm the schema is showing the attributes that you are interested in retaining. If not, you can click on **Edit** schema and make the necessary changes:

Schema

	Column name	Data type	Partition key
1	device_id	string	
2	timestamp	double	
3	device_name	string	
4	temperature	double	
5	humidity	double	
6	vibration	double	
7	duty_cycles	double	
8	partition_0	string	Partition (0)
9	partition_1	string	Partition (1)
10	partition_2	string	Partition (2)
11	partition_3	string	Partition (3)
12	partition_4	string	Partition (4)

Figure 6.9 – The table schema in Glue

7. Keep a note of the database and the table name, as you will need them in the next two sections.

In this lab, you used a crawler with a single data source only; however, you can add multiple data sources if required by your use case. Once the data catalog is updated and the data (or metadata) is available, you can consume it through different AWS services. You might need to often clean, filter, or transform your data as well. However, these responsibilities are not generally performed by the IoT practitioners and, primarily, fall with the data analysts or data scientists.

Amazon Athena

Amazon Athena acts as a serverless data warehouse where you run analytical queries on the data that's curated by an ETL engine such as Glue. Athena uses a schema-on-read approach; thus, a schema is projected onto your data when you run a query. And since Athena enables the decoupling of the compute and storage layers, you can connect to different data lake services such as S3 to run these queries on. Athena uses Apache Hive for DDL operations such as defining tables and creating databases. For the different functions supported through queries, Presto is used under the hood. Both Hive and Presto are open source SQL engines:

1. Navigate to the **AWS Athena** console, and choose **Data Sources** from the left-hand pane.

2. Keep the data source as **default** and choose the database name of `packt_gluedb`:

    -   This was created in the previous section by the Glue crawler automatically after scanning the S3 destination bucket, which is storing the batched sensor data.

3. This should populate the list of tables created under this database.

4. Click on the three dots adjacent to the table resembling the name of `*mysensordatabucket*` and select **Preview table**. This should automatically build and execute the SQL query.

This should bring up the data results with only 10 records. If you would like to view the entire dataset, please remove the 10-parameter limit from the end of the query. If you are familiar with SQL, please feel free to tweak the query and play with different attributes or join conditions.

> **Note**
>
> Here, you processed JSON data streamed from an HBS hub device. But what if your organization wants to leverage a more lightweight data format? Athena offers native support for various data formats such as CSV, AVRO, `Parquet`, and ORC through the use of **serializer-deserializer (SerDe)** libraries. Even complex schemas are supported through regular expressions.

So far, you have crawled the data from the data lake, created the tables, and successfully queried the data. Now, in the final step, let's learn how to build dashboards and charts that can enable BI on this data.

QuickSight

As an IoT practitioner, building business dashboards might not be part of your core responsibilities. However, some basic knowledge is always useful. If you think of traditional BI solutions, it might take data engineers weeks or months to build complex interactive ad hoc data exploration and visualization capabilities. Therefore, business users are constrained to pre-canned reports and preselected queries. Also, these traditional BI solutions require significant upfront investments and don't perform as well at scale as the size of data sources grow. That's where Amazon QuickSight helps. It's a managed service that's easy to use, highly scalable, and supports complex capabilities required for business:

1. Navigate to the Amazon QuickSight console and complete the one-time setup, as explained here:

    -   Enroll for the Standard Edition (if you have not used it before).

    -   Purchase SPICE capacity for the lab

    -   *Note that this has a 60-day trial, so be sure to cancel the subscription after the workshop to prevent getting charged.*

    -   Click on your login user (in the upper-right corner), and select **Manage QuickSight | Security & Permissions | Add and Remove | Check Amazon Athena | Apply**.

    -   Click on the QuickSight logo (in the upper-left corner) to navigate to the home page.

    -   Click on your login user (in the upper-right corner) and you will observe that your region preference is listed beneath your language preference.

    -   Confirm or update the region so that it matches your working region.

2. Click on **New Analysis**, then **New dataset**, and choose **Athena**.

3. Enter the data source name as `packt-data-visualization`, keep the workgroup as its default setting, and click on **Create Data Source**.

4. Keep the Catalog as **default**, choose **Database**, and then select the table created in *step 5* of the *AWS Glue* section.

5.   Click on **Select**, choose to directly query your data, and then click on **Visualize** again.

6.   Now build the dashboard:

   •   Choose a timestamp for the *X*-axis (select **MINUTE** from the **Value** drop-down menu).

   •   Choose the other readings such as **device_id**, **temperature**, and **humidity** for the *Y*-axis (select **Average** from the **Value** drop-down menu for each reading).

Feel free to play with different fields or visual types to visualize other smart home-related information. As you might have observed, while creating the dataset, QuickSight natively supports different AWS and third-party data sources such as Salesforce, ServiceNow, Adobe Analytics, Twitter, Jira, and more. Additionally, it allows instant access to the data through mobile apps (such as iOS and Android) for business users or operations to quickly infer data insights for a specific workload along with integrations to machine learning augmentation.

Congratulations! You have completed the entire life cycle of data processing and data consumption on the cloud using different AWS applications and data services. Now, let's wrap up this chapter with a summary and knowledge-check questions.

Challenge zone (Optional)

In the *Amazon API gateway* section, you built an interface to retrieve all the items from the `dynamodb` table. However, what if you need to extract a specific item (or set of items) for a particular device such as HVAC? That can be a less costly operation compared to scanning all data.

Hint: You need to define a route such as `GET /items {device_id}`. Check the lambda function to gain a better understanding of how it will map to the backend logic.

Summary

In this chapter, you were introduced to big data concepts relevant to IoT workloads. You learned how to design data flows using DDD approach along with different data storage and data integration patterns that are common with IoT workloads. You implemented a lambda architecture to process fleet telemetry data and an analytical pipeline. Finally, you validated the workflow by consuming data through the APIs and visualizing it through business dashboards. In the next chapter, you will learn how all of this data can be used to build, train, and deploy machine learning models.

Knowledge check

Before moving on to the next chapter, test your knowledge by answering these questions. The answers can be found at the end of the book:

1. Can you think of at least two benefits of domain-driven design from the standpoint of edge workloads?

2. True or false: bounded context and ubiquitous language are the same.

3. What do you think is necessary to have an operational datastore or a data lake/data warehouse?

4. Can you recall the design pattern name that brings together streaming and batch workflows?

5. What strategy could you incorporate to transform raw data on the cloud?

6. True or false: You cannot access data from a NoSQL data store through APIs.

7. When would you use a mediator versus broker topology for the event-driven workload?

8. Can you think of at least one benefit of using a serverless function for processing IoT data?

9. What **business intelligence (BI)** services can you use for data exposition to end consumers?

10. True or false: JSON is the most optimized data format for big data processing on the cloud.

11. How would you build an API interface on top of your operational data store (or data lake)?

References

Take a look at the following resources for additional information on the concepts discussed in this chapter:

-   *Data Management – Body of Knowledge*: `https://www.dama.org/cpages/body-of-knowledge`

-   *Domain-Driven Design* by Eric Evans: `https://www.amazon.com/Domain-Driven-Design-Tackling-Complexity-Software-ebook/dp/B00794TAUG`

-   *Domain Language*: `https://www.domainlanguage.com/ddd`

-   *Big Data on AWS*: `https://aws.amazon.com/big-data/use-cases/`

- *AWS Serverless Data Analytics Pipeline*: `https://d1.awsstatic.com/whitepapers/aws-serverless-data-analytics-pipeline.pdf`

- Modern serverless architecture on AWS: `https://d1.awsstatic.com/architecture-diagrams/ArchitectureDiagrams/mobile-web-serverless-RA.pdf?did=wp_card&trk=wp_card`

- *BI Tools*: `https://aws.amazon.com/blogs/big-data/tag/bi-tools/`

7
Machine Learning Workloads at the Edge

The growth of edge computing is not only driven by advancements in computationally efficient hardware devices but also by the advent of different software technologies that were only available on the cloud (or on-premises infrastructure) a decade back. For example, think of smartphones, smartwatches, or personal assistants such as **Amazon Alexa** that bring a mix of powerful hardware and software capabilities to consumers. Capabilities such as unlocking your phone or garage doors using facial recognition, having a conversation with Alexa using natural language, or riding an autonomous car have become the new normal. Thus, a need for cyber-physical systems to build intelligence throughout their lifetime based on continuous learning from their surroundings has become key for various workloads in today's world.

It's important to realize that most of the top technology companies (such as *Apple*, *Amazon*, *Google*, and *Meta*, formerly *Facebook*) use **machine learning** (**ML**) and have made it more accessible to consumers through their products. It's not a new technology, either, and has been used by industry sectors such as financial, healthcare, and industrial settings for a long time. In this chapter, we will focus on how ML capabilities can be leveraged for any **internet of things** (**IoT**) workload.

We will continue to work on the connected hub solution and learn how to develop ML capabilities on the **edge** (aka a **Greengrass** device). In the previous chapters, you have already learned about processing different types of data on the edge, and now, it's time to learn how different ML models can perform inference on this data to derive intelligent insights on the edge.

In this chapter, we will cover the following topics:

- Defining ML for IoT workloads
- Designing an ML workflow in the cloud
- Hands-on with ML workflow architecture

Technical requirements

The technical requirements for this chapter are the same as those outlined in *Chapter 2, Foundations of Edge Workloads*. See the full requirements in that chapter.

Defining ML for IoT workloads

ML technologies are no longer technologies of the future— they have transformed the lives of millions of people in the last few decades. So, what is ML?

> *"Machine Learning is the field of study that gives computers the ability to learn without being explicitly programmed."*
>
> *– Arthur Samuel, 1959*

Let's look at some real-world examples of ML applications for IoT workloads from the **consumer** and **industrial** segments.

First, here are some examples from the consumer segment:

- Smartphones or smartwatches that can identify your daily habits and provide recommendations related to fitness or productivity

- Personal assistants (such as Alexa, Google, and Siri) that you can interact with in a natural way for different needs such as controlling your lights and **heating, ventilation, and air conditioning (HVAC)**

- Smart cameras that can monitor your surroundings and detect unexpected behaviors or threats

- Smart garages that can recognize your car through its visual attributes, license plates, or even drivers' faces

- Self-driving cars that can continuously become smarter in identifying driving patterns, objects, and pedestrians in traffic

Here are some examples from the industrial segment:

- Smart factories that enable better optimization of **overall equipment effectiveness (OEE)**

- Better worker safety and productivity in different industrial plants, warehouses, or working sites

- Real-time **quality control (QC)** using **computer vision (CV)** or audio to identify defects

- Improving supply chain to reduce waste and enhance customer experience such as 1 hour or 15 mins delivery (with drones) from Amazon.com

To build the aforementioned capabilities, customers have used different ML frameworks and algorithms. For the sake of brevity, we are not going to cover every ML framework that exists today. We believe it's an area of data science that doesn't fit into the daily responsibilities of an IoT practitioner. But if you are interested in diving deeper, there are many books on ML/**artificial intelligence (AI)** available. Thus, our focus in this chapter will be to learn a bit about the history and core concepts of ML systems, and the approach to integrating ML with IoT and edge workloads.

What is the history of ML?

Today, as humans, we can communicate with machines of different kinds (from mobiles to self-driving cars) using voice, vision, or touch. This would not have been possible without the adoption of ML technologies. This is just the beginning, and ML will be more accessible in the years to come to transform our lives in different ways. You can take a quick tour through the history of this amazing technology in this article published by *Forbes*: `https://www.forbes.com/sites/bernardmarr/2016/02/19/a-short-history-of-machine-learning-every-manager-should-read/?sh=1ca6cea115e7`.

Beyond the research community, technology companies such as Amazon.com, Google, and others also started adopting ML technologies in the late 1990s. For example, Amazon.com used ML algorithms to learn about the reading preferences of their customers and built a model to notify them of new book releases matching their interests or genres. Google used ML for their search engine, Microsoft used it for identifying spam in emails, and so on. Since then, this technology has been adopted by many other industries for a plethora of use cases.

Now that we have learned a bit about the background of ML, let's now try to understand the foundation of ML.

What are the different types of ML systems?

Similar to distributed data systems, where there are different kinds of technologies to process different types of data, ML systems also come in different flavors. If we classify them into broader categories, the distinction can be described in this way:

- **Supervised ML (SML)**—In this method of ML, the model is trained with a labeled dataset and requires human supervision (or a teacher). For example, let's consider a scenario where we need a connected hub solution to identify different objects such as cats, dogs, humans (or seasonal birds?) who might have intruded onto your premises. Thus, the images are required to be labeled by a human (or humans), and the models will be trained on that data using a classification algorithm before they are ready (that is, the models) to predict the outcomes. In the following screenshot, you can see some objects are already labeled while the rest are not. So, humans are required to do their due diligence with labeling for the model to be effective:

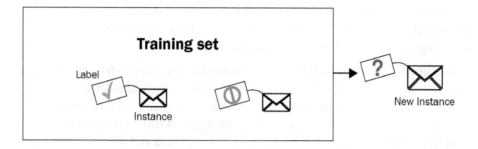

Figure 7.1 – A labeled training set of image classification

The length of the training and the volume and quality of the data will determine the accuracy of the model.

• **Unsupervised ML (UML)**—In this method of ML, the model is trained with an unlabeled dataset and requires no human supervision (or self-teaching). For example, let's consider that you have a new intruder on your premises, such as a deer, a wolf (or a tiger?), and you expect the model to detect that as an anomaly and notify you. In the following screenshot, you can see that none of the images is labeled and the model is required to figure out the anomaly on its own:

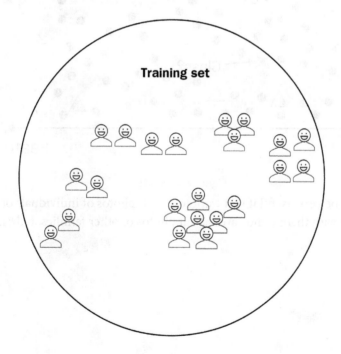

Figure 7.2 – An unlabeled training set of image classification

Considering the training dataset did not have pictures of a deer, a wolf, or a tiger, the model needs to be smart enough to identify that as an **anomaly** (or a novelty detection) using algorithms such as **random forest**.

- **Semi-supervised ML (SSML)**—In this method of ML, the model is trained with a dataset that is a mix of unlabeled and labeled data (think of on-demand teaching where you need to learn most of the content on your own). So, let's consider a scenario where you collected pictures of your guests from a party thrown at your home. Different guests show up in different photos and most of them are not labeled, which is the unsupervised part of the algorithm (such as **clustering**). Now, as the host of the party, if you label the unique individuals once in the dataset, the ML model can immediately recognize those individuals in other pictures on its own, as shown in the following diagram:

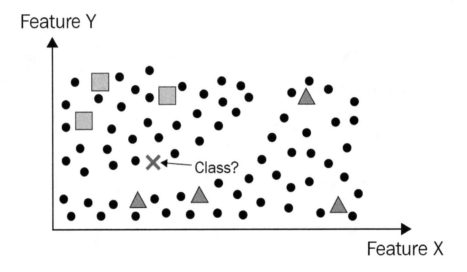

Figure 7.3 – SSML

This might be very useful if you want to search photos of individuals or families and share these with them (who cares about photos of other families, huh?).

- **Reinforcement learning (RL)**—In this method of ML, the model is trained to make a sequence of decisions in an environment and maximize a long-term objective. The model learns through an iterative process of trial and error. An agent such as a physical or virtual device uses this model to take actions guided by a policy at a given environment state and reaches a new state. This makes the agent eligible for a reward (positive or negative), and the agent continues to iterate on this process until it leads to the most optimal long-term rewards. The entire life cycle of an agent progressing from an initial state to a final state is called an **episode**.

For example, using RL, you can train a robot to take pictures of all guests from a party thrown at your home. The robot will stay on a specific track and capture images from that environment. It will receive a positive reward if it stays on track and captures acceptable images and a negative reward for going off track or capturing distorted snaps. As it continues to iterate through this process, eventually it will learn how to maximize its long-term objectives of capturing glorious images. The following diagram reflects this process of RL:

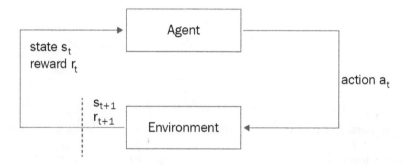

Figure 7.4 – RL

We have explained a broader category of ML in the preceding diagram; however, there is a plethora of frameworks and algorithms that are beyond the scope of this book. So, in the next section, we will focus on the most common ones that can apply to data generated from IoT and edge workloads.

Taxonomy of ML with IoT workloads

The three most common uses of ML in IoT are in the fields of classification, regression, and clustering, as depicted in the following diagram:

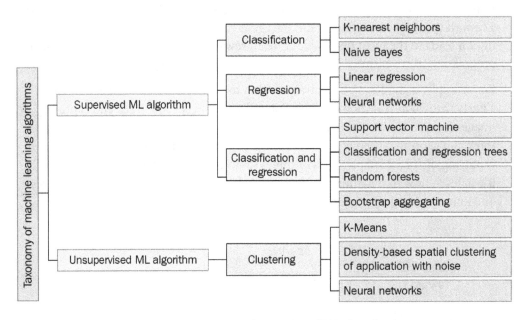

Figure 7.5 – Summarized taxonomy of ML algorithms

Let's discuss the nomenclature in the preceding diagram in more detail, as follows:

- **Classification**—Classification is an SML technique where you start with a set of observed values and derive some conclusion about unknown data. In the real world, classification can be used in image classification, speech recognition, drugs classification, sentiment analysis, biometric identification, and more.

- **Regression**—Regression is an SML technique where you can predict a continuous value. The prediction happens by estimating the relationship between the dependent variable (Y) and one or more independent variables (X) using a best-fit straight line. In the real world, regression can be applied to forecasting the temperature tomorrow, the price of energy utilization, the price of gold, and so on.

- **Clustering**—Clustering is a UML algorithm where you can group unlabeled data points. This is used very often for statistical analysis. The grouping of unlabeled data points in a dataset is performed by identifying data points in a dataset that share common properties and features. In the real world, this algorithm can be applied to market segmentation, medical imaging, anomaly detection, and social network analysis.

In the hands-on lab section of this chapter, you will learn how to use a classification algorithm to classify objects (such as cars and pets).

Why is ML accessible at the edge today?

We have already introduced you to three laws of edge computing in *Chapter 6, Processing and Consuming Data on the Cloud: Law of Physics* (latency-sensitive use cases), *Law of Economics* (cost-sensitive use cases), and *Law of the Land* (data-sensitive use cases). Based on these laws, we can identify various use cases in today's world, especially related to IoT and the edge, where it makes a lot of sense to process and generate insights from the data locally on the device or the gateway itself rather than publishing them continuously to the cloud.

However, the constraint is the limited resources (such as **central processing unit (CPU)**, **graphics processing unit (GPU)**, memory, network, and energy) available on these edge devices or gateways. Thus, it is recommended to take advantage of the computing power of the cloud to build and train ML models using the preferred framework (such as **MXNET**, **TensorFlow**, **PyTorch**, **Caffe**, or **Gluon**) and then deploy the model to the edge for inferencing.

For example, if there is a lot of noisy data generated in a smart home from a baby crying, a dog barking, or construction noises from the surroundings, the ML model can identify that as noisy data, trigger any specified action locally—such as an alert to check on the baby or the pet—but avoid publishing those data points to the cloud. In that way, a large amount of intermittent data that's of less long-term value can be filtered out at the site itself.

ML at the edge is an evolving space and there are many emerging frameworks and hardware offerings available today from different vendors, a few of which are listed in the following tables.

Here are common ML frameworks for the edge:

Framework	Developer	System requirement
TensorFlow Lite	Google	Android, iOS, microcontrollers
PyTorch	Facebook	Android, iOS
Core ML 3	Apple	iOS
MXNET	Apache	Linux, microprocessors
Embedded Learning Library (ELL)	Microsoft	Linux, microprocessors

Figure 7.6 – Common ML frameworks for the edge

Here are common hardware stacks for performing ML at the edge:

Hardware	Graphics processing unit (GPU)	CPU	Commonly used frameworks
AWS DeepLens	Generation 9 (Gen9) graphics	Intel Atom	Apache MXNet, TensorFlow, Caffe
AWS Panorama	NVIDIA Jetson Xavier	8-core Advanced RISC Machines (ARM) v8.2 64-bit CPU	Apache MXNet, TensorFlow, Keras, Darknet
Raspberry Pi 3, 4	Broadcom VideoCore VI	ARMv8-A	Apache MXNet, TensorFlow, OpenCV, Google Coral
NVIDIA Jetson TX2	NVIDIA Pascal	ARM® Cortex®-A57 MPCore	Apache MXNet, TensorFlow, Caffe, PyTorch, MATLAB

Figure 7.7 – Common hardware stacks for performing ML at the edge

You are already using Raspberry Pi as the underlying hardware for the different labs in this book. In the *Hands-on with ML architecture* section, you will learn how to train Apache MXNET-based ML models in the cloud and deploy them at the edge for inferencing. With this background, let's discuss how to get started with building ML applications for the edge.

Designing an ML workflow in the cloud

ML is an **end-to-end** (E2E) iterative process consisting of multiple phases. As we explain the different phases throughout the rest of the book, we will align to the general guidelines provided by **Cross Industry Standard Process for Data Mining** (**CRISP-DM**) consortium. The CRISP-DM reference model was conceived in late 1996 by three pioneers of the emerging data mining market and continued to evolve through participation from multiple organizations and service suppliers across various industry segments. The following diagram shows the different phases of the CRISP-DM reference model:

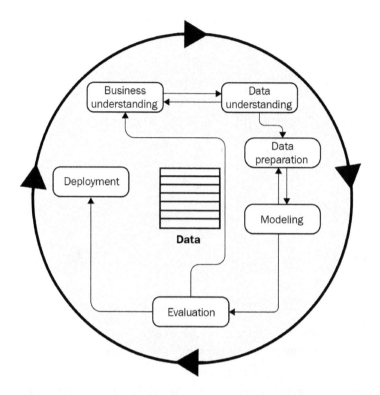

Figure 7.8 – Phases of the CRISP-DM reference model (redrawn from https://www.the-modeling-agency.com/crisp-dm.pdf)

This model is still considered a baseline and a proven tool for conducting successful data mining projects as its application is neutral and applies well to a wide variety of ML pipelines and workloads. Using the preceding reference model (*Figure 7.5*) as the foundation, the life cycle of an ML project can be expanded to the following activities:

Phases	Activities
Business understanding	Determining business objectives; framing ML problems
Data understanding	Data collection; integration; exploration; data quality inspection
Data preparation	Data cleaning; formatting; reconstructing; visualizing
Modeling	**Feature engineering (FE)**; model training
Evaluation	Assessment of results; model **key performance indicator (KPI)** reviews
Deployment	Plan deployment; monitoring and maintenance; business goal evaluation

Figure 7.9 – Life cycle of an ML project

The workflow of the preceding ML activities can be visually depicted as follows:

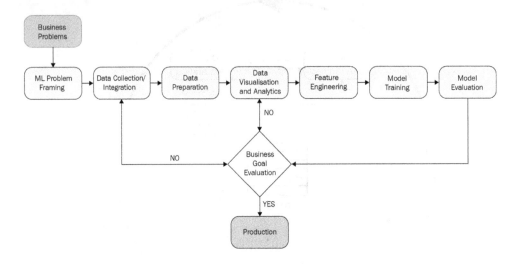

Figure 7.10 – E2E ML process

In the following section, we will elaborate on these concepts in detail by using an image classification scenario for your connected home.

Business understanding and problem framing

The first phase is working backward from the use case and understanding the requirements from a business perspective. Once that's clear, the business context gets translated to the technical requirements (such as the need for ML technologies) to achieve the required business outcomes. Does this concept sound familiar? If yes, congrats—you were able to relate to the concepts of **domain-driven design** (**DDD**), introduced in *Chapter 6, Processing and Consuming Data on the Cloud*. The ML capabilities can be treated as just another bounded context with its own set of ubiquitous languages.

But problem-solving using ML can be different, and here is a great quote from Peter Norvig (Director of Research at Google) on that:

> *"Machine Learning changes the way you think about a problem. The focus shifts from a mathematical science to a natural science, running experiments and using statistics, not logic, to analyze its results."*

An organization needs to clearly identify whether the business problem they are trying to solve is an ML problem. If the problem can be solved using traditional programming methods, building ML models might be overkill. For example, if you plan to forecast the future revenue of your business for a specific quarter based on historical data, traditional analytical methods might be sufficient. But if you start factoring prediction of other variables such as weather, competitor campaigns, promotions, economy, this becomes a better fit for an ML problem. So, as a rule of thumb, always try to start your ML journey with the following questions:

- *What problem is my organization or product facing?*

  Let's consider a scenario where you are trying to solve the problem of securing your family, pets, and neighbors from incoming traffic around your parking space.

- *Would it be a good problem to solve using ML or classic analytics methods?*

  Here, a connected **Home Base Solutions (HBS)** hub is required to identify any living things in the parking space when a vehicle approaches or departs the area. This problem cannot be solved using a classic analytics method because you won't have the schedule available for every visitor, neighbor, or delivery van for your area. Thus, the hub can detect the movement of a vehicle(s) using motion sensors, capture different images from the surroundings (using the installed camera), and run real-time inferences to detect any objects around it. If that's found, it will alert the driver, kids, or pets in real time and avoid accidents.

  Earlier, computer vision (CV) models relied on raw pixel data as an input, but this was found to be inefficient due to several other factors such as different backgrounds behind the object, lighting, camera angle, or focus. Thus, image classification is clearly an ML problem.

- *If it's an ML problem, do I have enough data of optimal quality?*

  Considering the objects in scope here are generic—such as humans, cats, or cars— we can rely on public datasets such as **Caltech-256**. This dataset contains more than 30,000 images of 256 different types of objects.

  This is often the most common question we come across: *How much data is enough for training?* It really depends.

You should have at least a few thousand data points for basic linear models, and hundreds of thousands for neural networks (such as with image classification, as previously mentioned). More data of optimal quality enables the model to predict smarter. If you have less data or data of poor quality, the recommendation is to consider a purpose-built AI service non-ML solution first. I often like to quote that with data, it's always *garbage in = garbage out*. Thus, if you have questionable data quality, it is of less value to a classic analytics method or an ML process. Additionally, an ML process is more expensive as you will waste a lot of time and resources and incur costs to train models with questionable performance.

Now, let's assume that you have met the preceding requirements and identified that the problem you are trying to solve is truly an ML one. In that case, you can choose the following best practices to summarize the problem framing:

1. Formulate the ML problem into a set of questions with its respective inputs and desired outputs.

2. Define tangible performance metrics for the project, such as accuracy, prediction, or latency.

3. Establish the definition of success for the project.

4. Frame a strategy for data sourcing and data annotation.

5. Start simple—build a model that is easy to interpret, test, debug, and manage.

Data collection or integration

In this of the E2E ML process, you will identify a dataset that will feed as an input to the ML pipeline and evaluate the appropriate means to collect that. In the previous chapters, you have learned that for different IoT use cases, AWS provides a number of ways to ingest the raw data in bulk or in real time. In other real-world scenarios, if you have **petabytes** (**PB**) of historical data in your cloud platform or data centers from IoT devices and **information technology** (**IT**) systems, there are multiple ways to transfer that to a data lake in the cloud, as follows:

*   Transfer over the public internet

*   Transfer over a private network using a dedicated fiber channel setup from your data centers to AWS using AWS Direct Connect

*   Transfer using hardware devices such as AWS Snowball, AWS Snowmobile, or AWS Snowcone, as it will take less time than transfering over the public internet

> **Fun Fact**
>
> Transporting data through snow devices is very similar to how you return a
> package to Amazon.com! You get hardware with an E-link screen acting as
> the return label, where the data can be loaded and shipped back to AWS data
> centers. If you have **exabytes (EB)** of data, AWS can even send you a truck
> for data transportation referred to as an AWS snowmobile. Please refer to the
> AWS documentation, *How to get started with AWS Snow Family* (`https://`
> `docs.aws.amazon.com/snowball/index.html`), to understand
> the required steps.

Again, consider the scenario of developing an ML model to identify vehicles approaching
a parking-space area. Here, you can use a training dataset from a public data repository
such as **Caltech**, as you mostly require generic images of kids, pets, and moving objects
such as cars and trucks to be classified. There will be two datasets in scope, as follows:

- **Training dataset**—Public dataset from Caltech to be hosted on a data lake
  (**Amazon Simple Storage Service (Amazon S3)**)

- **Inferencing dataset**—Generated in real time on the hub

The following code enables the downloading of two datasets from a public data repository:

```python
def download(url):
    filename = url.split("/")[-1]
    if not os.path.exists(filename):
        urllib.request.urlretrieve(url, filename)


def upload_to_s3(channel, file):
    s3 = boto3.resource("s3")
    data = open(file, "rb")
    key = channel + "/" + file
    s3.Bucket(bucket).put_object(Key=key, Body=data)


# caltech-256
download("http://data.mxnet.io/data/caltech-256/caltech-256-60-train.rec")
download("http://data.mxnet.io/data/caltech-256/caltech-256-60-val.rec")
```

Figure 7.11 – Data understanding

For a different use case where a public dataset is not an option, your organization needs to
have enough data points with optimal quality.

A summary of best practices for this phase is provided here:

- Define the various sources of data that you will use as an input to the ML pipeline

- Determine the form of data to be used as an input (that is, raw versus transformed) to the pipeline

- Use data lineage mechanisms to ensure that the data location and source are cataloged if required for further processing

- Use different AWS-managed services to collect, store, and process the data without additional heavy lifting

Data preparation

Data preparation is a key step in the pipeline as the ML models cannot perform optimally if the underlying data is not cleaned, curated, and validated. With IoT workloads, since the edge devices are co-existing in a physical environment with humans (over being hosted in a physical data center), the amount of noisy data generated can be substantial. In addition, as the dataset continues to grow from the connected ecosystem, data validation through schema comparisons can help detect if the data structure in newly obtained datasets has changed (for example, when a feature is deprecated). You can also detect if your data has started to drift—that is, the underlying statistics of the incoming data are different from the initial dataset used to train the model. Drift can happen due to an underlying trend or seasonality of the data or other factors.

Thus, the general recommendation is to start the data preparation with a small, statistically valid sample that can be iteratively improved with different strategies, such as the following:

- Checking for data anomalies

- Checking for changes in the data schema

- Checking for statistics of the different dataset versions

- Checking for data integrity, and so on

AWS provides a number of ways to help you prepare data at scale. You have already played with **AWS Glue** in *Chapter 5, Ingesting and Streaming Data from the Edge*. If you remember, AWS Glue allows you to manage the life cycle of data—such as to discover, clean, transform, and catalog. Once the data treatment is complete and the data quality meets the required standard, it can be fed as an input to an ML process.

In this chapter, we have introduced you to a different problem statement though, which is dealing with an unstructured dataset (aka images). Considering you are using a public dataset that's already labeled, you will only split the dataset into a training and a validation subset. The most common approach used by data scientists is to split the available data into a training and a test dataset, which is generally 70-30 (%) or 80-20 (%).

The following code enables the splitting of two datasets from a public data repository:

```
# Four channels: train, validation, train_lst, and validation_lst
s3train = "s3://{}/{}/train/".format(bucket, prefix)
s3validation = "s3://{}/{}/validation/".format(bucket, prefix)

# upload the lst files to train and validation channels
!aws s3 cp caltech-256-60-train.rec $s3train --quiet
!aws s3 cp caltech-256-60-val.rec $s3validation --quiet
```

Figure 7.12 – Data preparation

In the real world, though, you may not have clean or labeled data. Thus, you can leverage services such as **Amazon SageMaker Ground Truth**, which has an inbuilt capability to label data (such as images, text, audio, video) automatically along with easy access to public and private human labelers. This is useful if you lack in-house ML skills or are cost-sensitive to hiring data science professionals. Ground Truth uses an ML model to automatically label the raw data and produce high-quality training datasets at a fraction of a cost. But if the model fails to label the data confidently, it will route the problem to humans for resolution. Another aspect of data preparation is to understand the patterns in your dataset.

A summary of best practices for this phase is provided here:

- Profile your data through discovery and transformation.
- Choose the right tool for the right job (such as data labeling versus tuning).
- Understand the patterns from data compositions.

Data visualization and analytics

In this phase, you can continue the data exploration through various analytics and visualization tools to assess the data fitment for ML training post profiling. You can continue to leverage services such as Amazon Athena, Amazon Quicksight, and others introduced to you in *Chapter 6, Processing and Consuming Data on the Cloud*.

Feature engineering (FE)

In this phase, your responsibilities as IoT professionals are very limited. This is where the data scientists will determine the unique attributes in the dataset that can be useful in training the ML model. You can think of rows as observations and columns as properties (or attributes). As data scientists, your goal is to identify the columns that matter in solving a specific business problem (aka features). For example, with image classification, the color or brand of a car is not a key feature to determine it as a vehicle. This process of selecting and transforming variables to ensure the creation of an optimized ML model is referred to as **FE**. Thus, the key objective of FE is to curate data in a form that an ML algorithm can use to extract patterns and infer better results.

Let's break down the different phases of FE, as follows:

- **Feature creation** to identify the attributes from a dataset relevant to the problem in scope, such as height and width of the pixels for images
- **Feature transformation** for data compatibility or quality transformation, such as resizing inputs to a fixed size or converting non-numeric to numeric data
- **Feature extraction** to determine a reduced set of features that offers the most value
- **Feature selection** to filter redundant features from a dataset by observing variance of correlation thresholds

If the number of features in a dataset becomes substantially large compared to the observations it can generate, the ML model may suffer from a problem called **overfitting**. On the other hand, if the number of features is limited, the model may infer a lot of incorrect predictions. This problem is referred to as **underfitting**. In other words, the model has trained well on the test data but is unable to apply the generalization to new or unseen datasets. Thus, feature extraction can help optimize a set of features for ML processing that are sufficient to generate a comprehensive version of the original set. Other than reducing the overfitting risk, feature extraction also speeds up the training through data compression and accuracy improvements. Different feature extraction techniques include **Principal Component Analysis (PCA)**, **Independent Component Analysis (ICA)**, **Linear Discriminant Analysis (LDA)**, and **Canonical Correlation Analysis (CCA)**.

AWS provides a number of ways to help you perform FE on your dataset at scale in an iterative way. For example, Amazon SageMaker as a managed service provides a hosted **Jupyter** notebook environment where you can use scikit-learn libraries to perform FE. If your organization is already invested in an **extract, transform, load (ETL)** framework such as AWS Glue, **AWS Glue DataBrew**, or a managed Hadoop framework such as **Amazon Elastic MapReduce (Amazon EMR)**, the data scientists can perform FE and transformation there, prior to leveraging SageMaker to train and deploy the models.

Another option is using **Amazon SageMaker Processing**. This feature provides a fully managed environment for running analytics jobs for FE and model evaluation at scale, along with incorporating various security and compliance requirements.

Here is a summary of best practices for this phase:

- Evaluate the attributes from the dataset that fit the *feature* paradigm
- Consider features that are useful to solve the problem at hand and remove redundant ones
- Build an iteration mechanism to explore new features or feature combinations

Model training

The key activities in this phase include choosing an ML algorithm that's appropriate for your problem and then training the model with the preprocessed data (aka features) from the earlier phases. We have already introduced you to the ML algorithms that are most common for IoT workloads in the *Taxonomy of ML with IoT workloads* section. Let's dive a bit deeper into those algorithms, as follows:

- **Classification**—Classification can be applied in two ways; that is, binomial or multiclass. Binomial is useful when you have a set of observed values around two groups or categories, such as dog versus cat, or email being spam or not spam. Multiclass includes more than two groups or categories such as a set of flowers — roses, lilies, orchids, or tulips. Different classification techniques include decision trees, random forests, logistic regression, and naive Bayes.

- **Regression**—Classification is used to predict a discrete value, whereas regression is used to predict a continuous variable. Regression can be applied in three ways: *least square method, linear,* or *logistic.*

- **Clustering**—*K-means* is a very popular clustering algorithm generally used to assign a group to unlabeled data. This algorithm is fast and scalable as it uses a methodology to assign each group by computing the distance between the data point and each group center.

In the safety scenario around the parking space cited earlier, we are using a multiclass classification algorithm, since we expect the model to classify multiple categories of objects, such as humans (specially kids), cars, and animals (such as cats, dogs, and rabbits). AWS services such as SageMaker do the undifferentiated heavy lifting of creating and managing the underlying infrastructure required for the training. You can choose different types of instances, such as CPU- or GPU-enabled. In the following example, you only specify an instance type of `ml.p2.xlarge` along with other required parameters such as `volume` and `instance_count`, and SageMaker does the rest, using the estimator interface for instantiating and managing the infrastructure:

```
s3_output_location = "s3://{}/{}/output".format(bucket, prefix)
ic = sagemaker.estimator.Estimator(
    training_image,
    role,
    instance_count=1,
    instance_type="ml.p2.xlarge",
    volume_size=50,
    max_run=360000,
    input_mode="File",
    output_path=s3_output_location,
    sagemaker_session=sess,
)
```

Figure 7.13 – ML training infrastructure

You will be using the **MXNet framework** in this chapter, but SageMaker allows most other ML frameworks, such as TensorFlow, PyTorch, and Gluon, to train your model.

Please note that model optimization is a critical aspect of ML where you need to train a model with different sets of parameters to identify the most performant one. SageMaker hyperparameter tuning jobs help to optimize the models using Bayesian optimization or random search techniques. As you can see in the following example, the model is getting trained using hyperparameters such as batch size and shape to solve your business problem:

```
ic.set_hyperparameters(
    num_layers=18,
    image_shape="3,224,224",
    num_classes=257,
    num_training_samples=15420,
    mini_batch_size=128,
    epochs=5,
    learning_rate=0.01,
    top_k=2,
    precision_dtype="float32",
)
```

Figure 7.14 – ML training parameters

To make this process of model training easier and cost-effective for organizations new to ML, SageMaker supports automatic model tuning (through **Autopilot**) to automatically perform these actions on your behalf. It's also possible to use your custom ML algorithm as a container image and train it using SageMaker. For example, if you already have a homegrown image classification model that doesn't use any of the SageMaker-supported ML frameworks, you can use your model as a container image and retrain it in SageMaker without starting from scratch. SageMaker also offers a monitoring and debugging capability that allows clear visibility to the training metrics.

Here is a summary of best practices of this phase:

- Choose the right algorithm and training parameters for your data or let the managed services choose these for you

- Ensure the dataset is segregated into training and test sets

- Apply incremental learning to build the most optimized model

- Monitor the training metrics to ensure the model performance doesn't degrade over time

Model evaluation and deployment

In this phase, the model is evaluated to assess if it solves the business problem in context. If it doesn't, you can build multiple models with different business rules or methodologies (such as a different algorithm, other training parameters, and so on) until you find the optimized model that meets the business KPIs. Data scientists may often uncover inferences for other business problems in this phase as they test the model(s) against a real application. In order to evaluate the model, it can be tested against *historical data* (aka offline evaluation) or *live data* (aka online evaluation). Once the ML algorithm passes the evaluation, the next step is to deploy the model(s) to production.

The scenario for IoT/edge workloads gets a bit tricky, though, if your use cases primarily deal with offline processing and inferencing on the edge. In that case, you don't have access to the scale of the cloud, and thus the best practice is often to further optimize the models. **SageMaker Neo** can be useful in this scenario, as it allows you to train your model once and run anywhere in the cloud or at the edge. With this service, you can compile the model in most common frameworks (such as MXNET, TensorFlow, PyTorch, Keras) and deploy the optimized version on a target platform of your choice (such as hardware from Intel, NXP, NVIDIA, Apple, and so on). The following code helps to optimize the model based on different parameters such as OS and Architecture:

```python
response = client.create_compilation_job(
    CompilationJobName='<update-job-name-here>',
    RoleArn=role,
    InputConfig={
        'S3Uri': '<replace-with-s3-uri>',
        'DataInputConfig': '{"data": [1, 3, 224, 224]}',
        'Framework': 'MXNET'
    },
    OutputConfig={
        'S3OutputLocation': '<replace-with-s3-output-location>',
        'TargetPlatform': {
            'Os': 'LINUX',
            'Arch': 'ARM64'
        },
    },
    StoppingCondition={
        'MaxRuntimeInSeconds': 900,
        'MaxWaitTimeInSeconds': 900
    }
)
```

Figure 7.15 – Optimizing the model with SageMaker Neo

The way SageMaker Neo works is, its compiler uses an ML model under the hood to apply the best available performance for your model on the respective edge platform or device. Neo can optimize models to perform up to 25 times faster with no loss in accuracy and requires as little as one-tenth of the footprint compared to a non-optimized model. But how? Let's explore, as follows:

- Neo has a compiler and a runtime component.

- The Neo compiler has an **application programming interface (API)** that can read models developed using various frameworks. Once the read operation is complete, it converts the framework-specific functions and operations into a framework-agnostic intermediate representation.

- Once the conversion is complete, it then starts performing a set of optimizations. As a result of this optimization, binary code is generated and persisted to a shared object library, along with the model definition and parameters that are stored in separate files.

- Finally, Neo runtime APIs for the supported target platform can load and execute this compiled model to offer the required performance boost.

As you can imagine, this optimization is powerful for edge devices as they are resource-constrained. The following screenshot diagrammatically represents how you can deploy an optimized model in your production environment. In the *Hands-on with ML architecture* section of this chapter, you will learn how to deploy a model trained with SageMaker Neo:

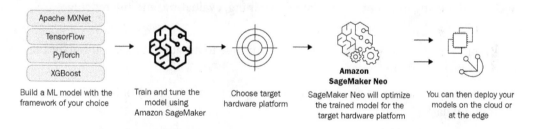

Figure 7.16 – Optimizing the ML model for the hardware

Once the model is deployed, you need to monitor the performance metrics over time. This is because it's pretty common for models to function less effectively as the real-world data may start to differ from the data that was used to train the model. The SageMaker Model Monitor service can help detect deviations and alert you, the data scientists, or ML operators, to take remedial action.

Here is a summary of best practices for this phase:

- Evaluate if the model performance meets the business goals
- Identify the inferencing method required for your models (offline versus online)
- Deploy the model on the cloud with automatic scaling options or on the edge with hardware-specific optimization
- Monitor the model performance in production to identify drift and perform remediations

ML design principles

Now that you have learned about the common activities in an ML workflow, let's summarize the design principles from the steps explained in the preceding section, as follows:

- Work backward from the use case to identify if it's a problem that needs ML or that can be solved using a classic analytical approach.
- Collect enough data of optimal quality to have accurate ML models.
- Perform profiling to understand the data relationships and compositions. Remember: *garbage in = garbage out*, thus data preparation is key in an ML process.
- Start with a small set of features (aka attributes) to solve a specific business problem and evolve through experimentation.
- Consider using different sets of data for training, evaluating, and inferencing purposes.
- Evaluate the accuracy of the models and continue to iterate until it's optimal for the business problem in context.
- Determine if the model needs to inference on real-time (online) or historical (offline) data.
- Host different variants of the model on the cloud or on the edge and identify the most optimal one against real-world data.
- Continuously monitor the metrics from the deployed models for accuracy, remediation, and improvement.

- Leverage managed services for offloading the heavy lifting of managing the underlying infrastructure.

- Automate the different activities of the pipeline as much as possible, such as data preparation, model training, evaluation, hosting, monitoring, and alerting.

ML anti-patterns for IoT workloads

Similar to any other distributed solutions, IoT workloads based on ML have anti-patterns as well. Here are some of them:

- **Don't put all your eggs in one basket**—The E2E ML process includes many activities and thus requires different personas such as the following:

  - *Data engineer*—For data preparation, designing ETL (or ELT) processes

  - *Data scientists*—For building, training, and optimizing the models

  - *Development-operations (DevOps or MLOps) engineers*—For building a scalable and repeatable machine learning infrastructure built with operating and monitoring mechanisms

  - *IoT engineers*—For building a cloud-to-edge deployment pipeline along with integration from the IoT gateway to different backend services

  In summary, expecting a single resource to perform all these activities at scale will lead to the failure of the project.

- **Don't assume the requirements; get aligned**—A plethora of tools is available for analytical and ML purposes and each has its pros and cons. For example, consider the following:

  - Both **R** and **Python** are popular programming languages for developing ML systems. In general, business analysts or statisticians will prefer R or other commercial solutions (such as **MATLAB**), while data scientists will choose Python.

  - Similarly, data scientists may have their preferred ML framework, such as TensorFlow over PyTorch. Choosing different languages or frameworks has downstream impacts—for example, the edge hardware may need to support the respective version or libraries required for inferencing the ML model.

  Thus, it's important for all the different personas engaged in ML activities to stay aligned on the business and technical requirements.

- **Plan for technical debt**—ML systems for IoT workloads are prone to accruing technical debt due to their data dependencies from IoT sources that are often unreliable due to their noisiness. This may also happen due to dependencies on other upstreams having inconsistent data. For example, consider the following:

  - If the composition of any feature (or a few) changes substantially, it will lead the model to behave differently in the real world. This problem is also referred to as the **training-serving skew**, where there is a discrepancy in how you handle data in the training and serving pipelines.

  - The reason ML workloads are different from traditional IT systems is their behavior can be determined only through real data over unit testing with a small sample.

Thus, it's key to monitor the accuracy of the model, optimize it iteratively based on the knowledge of the gathered data, and redeploy.

Hands-on with ML architecture

In this section, you will deploy a solution on a connected HBS hub that will require you to build and train ML models on the cloud and then deploy them to the edge for inferencing. The following screenshot shows the architecture of the lab with the highlighted steps *(1-5)* that you will complete:

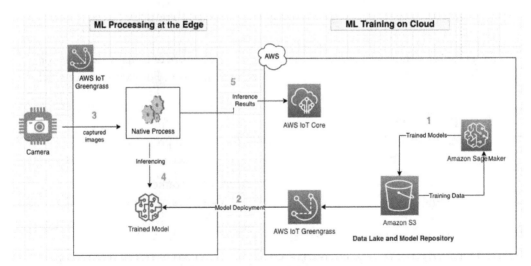

Figure 7.17 – Hands-on ML architecture

Your objectives include the following, which are highlighted as distinct steps in the preceding architecture:

- Build the ML workflow using Amazon SageMaker
- Deploy the ML model from the cloud to the edge using AWS IoT Greengrass
- Perform ML inferencing on the edge and visualize the results

The following table shows the list of components you will use during the lab:

Edge services	Cloud services
Greengrass Nucleus	AWS IoT Core
Greengrass Stream Manager	Amazon SageMaker
Greengrass ML Inference	Amazon S3
Native processes	

Figure 7.18 – Hands-on lab components

Building the ML workflow

In this section, you will build, train, and test the ML model using Amazon SageMaker Studio.

> **Note**
> Training models using Amazon SageMaker will incur additional cost. If you want to save on that, please use a trained ML model available in GitHub for your platform and skip to the next section, that is, Deploying the model from cloud to the edge.

Amazon Sagemaker Studio is a web-based **integrated development environment** (IDE) that enables data scientists (or ML engineers) with a single-stop shop for all things ML. To train the model, you will use a public dataset from Caltech that has a collection of over 30,000 images across 256 object categories. Let's begin. Proceed as follows:

1. Please navigate to the **Amazon SageMaker** console and select **SageMaker Domain Studio** (from the left pane). If this is the first time you are interacting with the studio, you will be prompted to complete a one-time setup. Please choose **Quick setup,** click **Submit,** choose **Default VPC with a subnet (s) of your choice**, then click **Save and continue.**

2. It will take a few minutes for the studio to be set up. Please wait until the status shows **Ready** and then click **Launch app** -> **Studio.**

3. This should open up the SageMaker Studio (aka Jupyter console) for you. In case you are new to Jupyter, consider this as an IDE for developing ML models similar to Eclipse, Visual Studio, and so on, used for developing distributed applications.

4. Please upload the Jupyter notebook (`Image-Classification*.ipynb`) and the `synset.txt` file from the `chapter7/notebook` folder using the **Upload file** button in the top-left pane.

5. Double-click to open the Jupyter notebook and choose the Python runtime and kernel, as shown in the following screenshot:

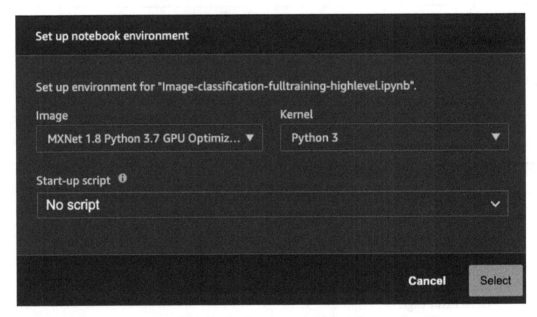

Figure 7.19 – Jupyter notebook kernel

6. Choosing a kernel is a critical step as it provides you with the appropriate runtime for training the ML model. Update the kernel (top right) to choose the GPU runtime for training the ML model. Since you will be processing images, a GPU runtime is preferable. After choosing the kernel, the Jupyter notebook will have the following kernel and configuration:

Figure 7.20 – Choosing a kernel

7. Now, navigate through the code slowly. Please ensure you read the text preceding the code to better understand the functioning of each of these blocks.

8. Click the **Run** button, as highlighted in the following screenshot, to execute one block at a time. Please don't click on the **Run** button if you see an asterisk (*) adjacent to the block. The asterisk implies that the code is still running. Please wait for that to disappear before you proceed:

Figure 7.21 – Running the steps

9. If you go through all the steps till the end, you will be able to complete the following steps:

    I.   Downloading a training dataset

    II.   Preparing and preprocessing the data

    III.   Splitting the dataset into training and test samples

    IV.   Training the model using the appropriate framework and parameters

V.   Optimizing the model for the edge hardware

VI. Hosting the model artifacts on the Amazon S3 repository

The model training can take up to 10 minutes to complete. After the training, you have an ML model that's trained using the MXNet framework and is capable of performing classification on 256 different objects.

10. Please navigate through the S3 bucket (`sagemaker-<region>-<accountid>/` `ic-fulltraining/`). Copy the S3 **Uniform Resource Identifier** (**URI**) for the model object (`DLR-resnet50-*-cpu-ImageClassification.zip`), as you will need it in the next section.

Now that the model is trained on the cloud, you will deploy it on the edge using Greengrass for near-real-time inferencing.

Deploying the model from cloud to the edge

As an IoT practitioner, deploying the model from the cloud to the edge is the step you will primarily be involved in. This is the transition point, where the ML or data science team provides a model that needs to be pushed to a fleet of devices on the edge. Although it's possible to automate these steps in the real world using a **continuous integration/continuous deployment** (**CI/CD**) pipeline, you will do it manually to learn the process in detail.

Similar to the components you have created in the earlier chapters for deploying different processes on the edge (such as publisher, subscriber, aggregator), deploying an ML resource needs the same approach. We will create a component that includes the ML model trained in the previous section. We will continue to use the AWS console for continuity.

> **Note**
>
> Greengrass provides a sample image classification model component that you toyed with in *Chapter 4*, *Extending the Cloud to the Edge*. Here, you are learning how to modify existing model components with your custom resources. In the real world, you may even have to create a new model component from scratch, where you can follow a similar process.

Let's begin. Proceed as follows:

1. Please navigate to the **Amazon IoT Greengrass** console, click on **Components**, click on the **My components** tab, and then on **Create Component**. Under **Component information**, select **Enter recipe as JSON** as your component source.

In the **Recipe** box, paste the component recipe from the `chapter7/recipe` folder. Now, let's update the recipe to point to the trained model. Please replace the URI (marked with an arrow) with the S3 URI copied in *Step 10* of the *Building the ML workflow* section, as illustrated in the following screenshot. If you have skipped the earlier section and used a trained model from GitHub, please manually upload the model to the S3 bucket of your choice and update the S3 URI in the recipe accordingly:

```
{
    "RecipeFormatVersion": "2020-01-25",
    "ComponentName": "variant.DLR.ImageClassification.ModelStore",
    "ComponentVersion": "2.1.6",
    "ComponentType": "aws.greengrass.generic",
    "ComponentDescription": "Downloads the resnet50_v1 image classification ML
        models to the device as artifacts.",
    "ComponentPublisher": "AWS",
    "ComponentDependencies": {
        "aws.greengrass.Nucleus": {
            "VersionRequirement": ">=2.0.0 <2.6.0",
            "DependencyType": "SOFT"
        }
    },
    "Manifests": [
        {
            "Platform": {
                "os": "linux",
                "architecture": "aarch64"
            },
            "Lifecycle": {},
            "Artifacts": [
                {
                    "Uri": "s3://sagemaker-<region>-<account-id>/ic-fulltraining
                        /DLR-resnet50-aarch64-cpu-ImageClassification.zip",
                    "Algorithm": "SHA-256",
                    "Unarchive": "ZIP",
                    "Permission": {
                        "Read": "OWNER",
                        "Execute": "NONE"
                    }
                }
            ]
        }
    ],
    "Lifecycle": {}
}
```

Figure 7.22 – Recipe configuration

2. Click **Create component** to finish creating the model resource. This model should appear on the **My components** tab of the **Components** page.

3. Now, you need to create an inferencing component. This is the resource that triggers the model on the edge and publishes the results to the cloud.

> **Note**
>
> Greengrass provides a sample image classification inferencing component, `aws.greengrass.DLRImageClassification`, that you already played with in *Chapter 4, Extending the Cloud to the Edge*. Here, you are learning how to use the same inferencing component to work with your custom ML model. In the real world, you may have to create a new inferencing component from scratch with a modified manifest file, as you just did with the model.

4. On the **Amazon Greengrass** console, choose **Deployments**. On the **Deployments** page, revise your existing deployment and choose the following components:

    - `variant.DLR.ImageClassification.ModelStore`: ML model trained through SageMaker. You can choose this from the **My components** tab.

    - `aws.greengrass.DLRImageClassification`: Inferencing script for the ML model. You can choose this from the **Public Component** tab.

    - `variant.DLR`: Runtime required for the ML inferencing. You can choose this from the **Public Component** tab.

5. On the **Select Components** page, ensure the components shown in the following screenshot are chosen, and then click **Next**:

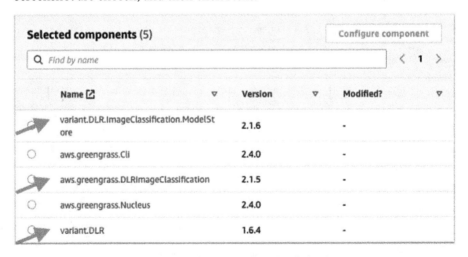

Figure 7.23 – Greengrass component dependencies

6. On the **Configure components** page, select the `aws.greengrass.`
    `DLRImageClassification` component. This will enable the **Configure**
    component option; click on that.

7. Update the **Configuration to merge** section (right pane) with the following
    configuration and click on **Confirm**:

```
{
    "InferenceInterval": "5"
}
```

8. Keep all the other options as default in the following screens and choose **Deploy**.

Let's now move on to the next section.

Performing ML inferencing on the edge and validating results

So, by now, the models are built, trained, and deployed on Greengrass. The images to be
inferred are stored in the following directory:

```
root@ubuntu-linux-20-04-desktop:/greengrass/v2/packages/artifacts-unarchived/aws.greengrass.DL
RImageClassification/2.1.6/image_classification/sample_images# ls -rlt
total 276
-r--r----- 1 ggc_user ggc_group 150656 Nov 23 16:01 cat.npy
-r--r----- 1 ggc_user ggc_group 127848 Nov 23 16:01 cat.jpeg
```

Figure 7.24 – Images directory on Greengrass hub

Similar to the cat image, you can deploy other images as well (such as dogs, humans, and
so on) and configure the inferencing component to infer on them.

Let's visualize the inferencing results that are being published from the edge to the cloud.
These results are key in assessing if the model is performing at the expected accuracy level.
We'll proceed as follows:

1. First, let's check if the component status shows as running and also check the
    Greengrass log to verify that there are no exceptions. Here's the code we'll need to
    do this:

```
sudo /greengrass/v2/bin/greengrass-cli component list
sudo tail -100f /greengrass/v2/logs/greengrass.log
sudo tail -100f aws.greengrass.DLRImageClassification.log
```

2. If there are no errors, please navigate to the AWS IoT console, choose **Test**, and then choose **MQTT test client**.

3. Under **Subscriptions**, choose `ml/dlr/image-classification`. Click **Subscribe** to view the results, which should look like this:

▼ ml/dlr/image-classification

```
{
  "timestamp": "2021-11-23 18:03:52.519163",
  "inference-type": "image-classification",
  "inference-description": "Top 5 predictions with score 0.3 or above ",
  "inference-results": [
    {
      "Label": "chambered nautilus, pearly nautilus, nautilus",
      "Score": "0.330684"
    }
  ]
}
```

Figure 7.25 – Inferenced results from ML

If you get stuck or require help, please refer to the **Troubleshooting** section in GitHub.

Congratulations on finishing the hands-on section of this chapter! Now, your connected **HBS hub** is equipped with ML capabilities that can operate in both online and offline conditions and can classify humans, pets, and vehicles in your parking-space area.

> **Challenge Zone (Optional)**
>
> Can you figure out how to act on the inference results published to AWS IoT Core? This will be useful to automatically trigger an announcement through notifications/alarms for cautioning kids/pets strolling in the parking-space area.
>
> **Hint**: You need to define a routing logic in the IoT Rules Engine to push the results through a notification service.

Isn't it incredible that you have now learned how to build ML capabilities for IoT workloads on the edge? It's time for a well-deserved break! Let's wrap up this chapter with a quick summary and a set of questions.

Summary

In this chapter, you were introduced to ML concepts relevant to IoT workloads. You learned how to design ML pipelines, along with optimizing models for IoT workloads. You implemented an edge-to-cloud architecture to perform inferences on unstructured data (images). Finally, you validated the workflow by visualizing the inferencing results from the edge for additional insights.

In the next chapter, you will learn how to implement DevOps and MLOps practices to achieve operational efficiency for IoT edge workloads deployed at scale.

Knowledge check

Before moving on to the next chapter, test your knowledge by answering these questions. The answers can be found at the end of the book:

1. True or false: Two types of ML algorithms exist: supervised and unsupervised.

2. Can you recall the four types of ML systems and their significance?

3. True or false: K-means is a classification algorithm.

4. Can you put the three phases of the ML project life cycle in the right order?

5. Can you think of at least two common frameworks used for training ML models?

6. What is the AWS service used for deploying trained models from the cloud to the edge?

7. True or False: AWS IoT Greengrass only supports custom components for image classification problems.

8. Can you tell me about one anti-pattern for ML with IoT workloads?

References

Take a look at the following resources for additional information on the concepts discussed in this chapter:

- CRISP: https://www.datascience-pm.com/crisp-dm-2/

- *A Short History of Machine Learning*: https://www.forbes.com/sites/bernardmarr/2016/02/19/a-short-history-of-machine-learning-every-manager-should-read/?sh=1ca6cea115e7

- *Machine Learning on AWS*: https://aws.amazon.com/machine-learning/

- ML with AWS IoT services: `https://aws.amazon.com/blogs/iot/category/artificial-intelligence/sagemaker/`

- *Using AWS IoT Greengrass Version 2 with Amazon SageMaker Neo and NVIDIA DeepStream Applications*: `https://aws.amazon.com/blogs/iot/using-aws-iot-greengrass-version-2-with-amazon-sagemaker-neo-and-nvidia-deepstream-applications/`

- Well-Architected Framework for ML: `https://docs.aws.amazon.com/wellarchitected/latest/machine-learning-lens/machine-learning-lens.html`

- Learn from AWS through **Machine Learning University (MLU)**: `https://aws.amazon.com/machine-learning/mlu/`

Section 3: Scaling It Up

In this section, you will learn how to provision IoT fleets at scale and use DevOps practices to build agile and operationally efficient IoT workloads at scale using different well-architected practices.

This section comprises the following chapters:

- *Chapter 8, DevOps and MLOps for the Edge*
- *Chapter 9, Fleet Management at Scale*

8
DevOps and MLOps for the Edge

The 21st century's flurry of connected devices has transformed the way we live. It can be hard to remember the days without the convenience of smartphones, smartwatches, personal digital assistants (such as Amazon Alexa), connected cars, smart thermostats, or other devices.

This adoption is not going to slow down anytime soon as the industry forecasts that there will be over 25 billion IoT devices globally in the next few years. With the increased adoption of connected technologies, the new normal is to have *always-on* devices. In other words, the *devices should work all the time.* Not only that, but we also expect these devices to continuously get smarter and stay secure throughout their life cycles with new features, enhancements, or bug fixes. But how do you make that happen reliably and at scale? Werner Vogels, Amazon's chief technology officer and vice president, often says that *"Everything fails all the time."* It's challenging to keep any technological solution up and running all the time.

With **IoT**, these challenges are elevated and more complicated as the **edge** devices are deployed in diverse operating conditions, exposed to environmental interferences, and have multiple layers of connectivity, communication, and latency. Thus, it's critical to build an edge-to-cloud continuum mechanism to collect feedback from the deployed fleet of edge devices and act on them quickly. This is where DevOps for IoT helps. **DevOps** is short for **development and operations**. It facilitates an agile approach to performing **continuous integration and continuous deployment (CI/CD)** from the cloud to the edge.

In this chapter, we will focus on how DevOps capabilities can be leveraged for IoT workloads. We will also expand our discussion to **MLOps** at the edge, which implies implementing agile practices for **machine learning (ML)** workloads. You learned about some of these concepts in the previous chapter when you built an ML pipeline. The focus of this chapter will be on deploying and operating those models efficiently.

You are already familiar with developing local processes on the edge or deploying components from the cloud in a decoupled way. In this chapter, we will explain how to stitch those pieces together using DevOps principles that will help automate the development, integration, and deployment workflow for a fleet of edge devices. This will allow you to efficiently operate an intelligent distributed architecture on the edge (that is, a Greengrass-enabled device) and help your organization achieve a faster time to market for rolling out different products and features.

In this chapter, we will be covering the following topics:

- Defining DevOps for IoT workloads
- Performing MLOps at the edge
- Hands-on with deploying containers at the edge
- Checking your knowledge

Technical requirements

- The technical requirements for this chapter are the same as those outlined in *Chapter 2, Foundations of Edge Workloads*. See the full requirements in that chapter.

Now, let's dive into this chapter.

Defining DevOps for IoT workloads

DevOps has transformed the way companies do business in today's world. Companies such as Amazon, Netflix, Google, and Facebook conduct hundreds or more deployments every week to push different features, enhancements, or bug fixes. The deployments themselves are typically transparent to the end customers in that they don't experience any downtime from these constant deployments.

DevOps is a methodology that brings *developers and operations* closer to infer quantifiable technical and business benefits with faster time to market through shorter development cycles and increased release frequency. A common misunderstanding is that DevOps is only a set of new technologies to build and deliver software faster. DevOps also represents a cultural shift to promote ownership, collaboration, and cohesiveness across different teams to foster innovation across the organization. DevOps has been adopted by organizations and companies of all sizes for distributed workloads to deliver innovation, enhancements, and operational efficiency faster. The following diagram shows the virtuous cycle of software delivery:

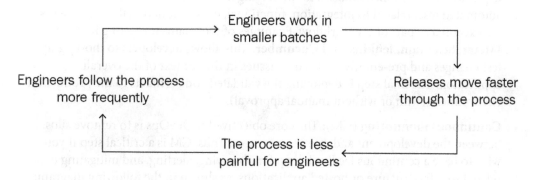

The Virtuous Circle of Software Delivery

Figure 8.1 – The virtuous cycle of software delivery

For the sake of brevity, we are not going to dive deeper into the concepts of DevOps or **Agile** practices here. Instead, we will focus on introducing the high-level concepts surrounding DevOps and discuss its relevance for IoT workloads.

Fundamentals of DevOps

DevOps brings together different tools and best practices, as follows:

- **Shared code repository**: Using a version control system is a prerequisite and a best practice in the field of code development. All artifacts that are required in the deployment package need to be stored here. Examples include **Bitbucket**, **Gitlab**, and **AWS CodeCommit**.

- **Continuous integration** (**CI**): In this step, developers commit their code changes regularly in the code repository. Every revision that is committed will trigger an automated build process that performs code scanning, code reviews, compilation, and automated unit testing. This allows developers to identify and fix bugs quickly, allowing them to adhere to the best practices and deliver features faster. The output of this process includes build artifacts (such as binaries or executable programs) that comply with the organization's enforced practices. Examples of toolchains include **Jenkins**, **Bamboo**, **GitLab CI**, and **AWS CodePipeline**. For IoT workloads, similar toolchains can be used.

- **Continuous delivery** (**CD**): This step expands on the previous step of CI and deploys all the compiled binaries to the staging or test environment. Once deployed, automated tests related to integration, functional, or non-functional requirements are executed as part of the workflow. Examples of toolchains for testing include **JMeter**, **Selenium**, **Jenkins**, and **Cucumber**. This allows developers to thoroughly test changes and pre-emptively discover issues in the context of the overall application. The final step is deploying the validated code artifacts to the production environment (with or without manual approval).

- **Continuous monitoring** (**CM**): The core objective for DevOps is to remove silos between the development and operations teams. Thus, CM is a critical step if you wish to have a continuous feedback loop for observing, alerting, and mitigating issues related to infrastructure or hosted applications, as shown in the following diagram:

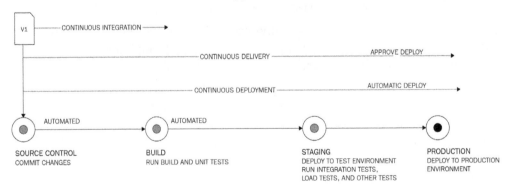

Figure 8.2 – DevOps life cycle

Common toolchains for monitoring include **Amazon CloudWatch**, **Amazon X-Ray**, **Splunk**, and **New Relic**.

- **Infrastructure as Code (IaC)**: Adhering to the software development practices of CI/CD to expedite shipping code is a great first step, but it's not enough. Teams can develop and test their code using agile processes, but the final delivery to production still follows waterfall methods. This is often due to a lack of control regarding provisioning or scaling the infrastructure dynamically. Traditionally, organizations will have system admins to provision the required infrastructure resources manually, which can take days, weeks, or months. This is where IaC helps as it allows you to provision and manage the infrastructure, configurations, and policies using code (or APIs) in an automated fashion without requiring any manual interventions that might be error-prone or time-consuming. Common toolchains include **Amazon CloudFormation**, **HashiCorp Terraform**, and **Ansible**.

Now that we have covered the the basics of DevOps, let's understand its relevance to IoT and the edge.

Relevance of DevOps for IoT and the edge

The evolution of edge computing from simple radio frequency identification systems to the microcontrollers and microprocessors of today has opened up different use cases across industry segments that require building a distributed architecture on the edge. For example, the connected HBS hub has a diverse set of functionalities, such as the following:

- A gateway for backend sensors/actuators
- Runtime for local components
- Interface to the cloud
- Message broker
- Datastream processor
- ML inferencing engine
- Container orchestrator

That's a lot of work on the edge! Thus, the traditional ways of developing and delivering embedded software are not sustainable anymore. So, let's discuss the core activities in the life cycle of an IoT device, as depicted in the following table, to understand the relevance of DevOps:

DevOps Concepts	Edge Activities	Cloud Activities
Device Manufacturing	• Supply chain • Device assembly	• Credentials provider such as a **Certificate Authority (CA)**
Device Registration	• Bootstrap	• Update registry • Activate credentials
Continuous Integration	Integration tests for the following: • Device firmware • App software	Build and test the following: • Device firmware • App software
Continuous Delivery	• Firmware installation • Software installation	• Deliver updates **over the air (OTA)**
Continuous Monitoring	• Log behavior • Measure performance • Alerts about errors	• Audit logs • Monitor key performance indicators • Act on alerts
Infrastructure as Code	N/A	• Provision and manage IT resources
End of Life	• Decommission hardware	• Deactivate credentials • Deregister devices

Figure 8.3 – Relevance of DevOps in IoT workloads

The key components of DevOps such as CI/CD/CM are equally relevant for IoT workloads. This set of activities is often referred to as **EdgeOps** and, as we observed earlier, they are applied differently between the edge and the cloud. For example, CI is different for the edge because we need to test device software on the same hardware that is deployed in the world. However, because of the higher costs and risks associated with edge deployments, it is common to reduce the frequency of updating devices at the edge. It is also common for organizations to have different sets of hardware for prototyping versus production runtimes.

DevOps challenges with IoT workloads

Now that you understand how to map DevOps phases to different IoT activities, let's expand on those a bit more. The following diagram shows the workflow that's typically involved in the life cycle of a device, from its creation to being decommissioned:

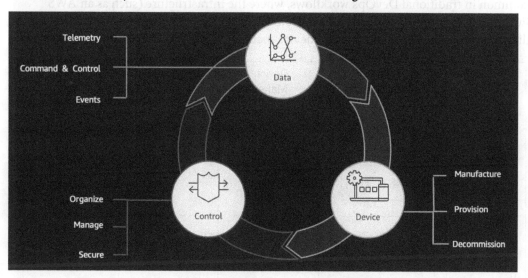

Figure 8.4 – DevOps workflow for IoT

Here, you can see some key differences between an IoT workload and other cloud-hosted workloads. Let's take a look.

The manufacturing process is involved:

Distributed workloads such as web apps, databases, and APIs use the underlying infrastructure provided by the cloud platform. Software developers can use IaC practices and integrate them with other CI/CD mechanisms to provision the cloud resources that are automatically required to host their workload. For edge workloads, the product lives beyond the boundaries of any data center. Although it's possible to run edge applications on virtual infrastructure provided by the cloud platform during the testing or prototyping phases, the real product is always hosted on hardware (such as a **Raspberry Pi** for this book's project). There is always a dependency on the contract manufacturer (or other vendors) in the supply chain for manufacturing the hardware, as per the required specifications that are followed for programming it with the device firmware. Although the firmware can be developed on the cloud using DevOps practices, flashing the firmware image is done at manufacturing time only. This hinders the end-to-end automation common in traditional DevOps workflows, where the infrastructure (such as an AWS EC2 instance) is readily imaged and available for application deployments. The following diagram shows the typical life cycle of device manufacturing and distribution:

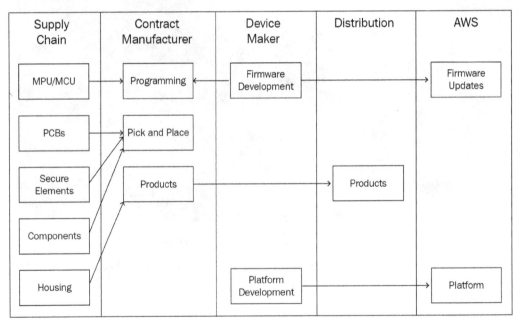

Figure 8.5 – IoT device manufacturing process

Securing the hardware is quintessential:

Some of the key vulnerabilities for edge workloads that are listed by **The Open Web Application Security Project (OWASP)** are as follows:

- Weak, guessable, or hardcoded passwords
- Lack of physical hardening
- Insecure data transfer and storage
- Insecure default settings
- Insecure ecosystem interfaces

Although distributed workloads may have similar challenges, mitigating them using cloud-native controls makes them easier to automate than IoT workloads. Using AWS as an example, all communications within AWS infrastructure (such as across data centers) are encrypted in transit by default and require no action. Data at rest can be encrypted with a one-click option (or automation) using the key management infrastructure provided by AWS (or customers can bring their own). Every service (or hosted workloads) needs to enable access controls for authentication and authorization through cloud-native **Identity & Access Management** services, which can be automated as well through IaC implementation. Every service (or hosted workload) can take advantage of observability and traceability through cloud-native monitoring services (such as **Amazon CloudTrail** or **Amazon CloudWatch**).

On the contrary, for edge workloads, all of the preceding requirements are required to be fulfilled during manufacturing, assembling, and registering the device, thus putting more onus on the supply chain to manually implement these over one-click or automated workflows. For example, as a best practice, edge devices should perform mutual authentication over TLS1.2 with the cloud using credentials such as X.509 certificates compared to using usernames and passwords or symmetric credentials. In addition, the credentials should have least-privileged access implemented using the right set of permissions (through policies). This can help ensure that the devices are implementing the required access controls to protect the device's identity and that the data in transit is fully encrypted. In addition, device credentials (such as X.509 certificates) on the edge must reside inside a secure element or **trusted platform module (TPM)** to reduce the risk of unauthorized access and identity compromise. Additionally, secure mechanisms are required to separate the filesystems on the device and encrypt the data at rest using different cryptographic utilities such as **dm-crypt**, **GPG**, and **Bitlocker**. Observability and traceability implementations for different edge components are left to the respective owners.

Lack of standardized frameworks for the edge:

Edge components are no longer limited to routers, switches, miniature servers, or workstations. Instead, the industry is moving toward building distributed architectures on the edge in different ways, as follows:

- **Fog computing**, which lets us shift more intelligence to the edge using a decentralized computing infrastructure of heterogeneous nodes

- **Mobile/Multi-Access Computing (MEC)**, which incorporates next-generation radio spectrums (such as 5G) to enable a new generation of workloads possible for the edge

- **Data center-in-a-box**, which enables resource-intensive computing capabilities at the edge with integrations to the cloud

The following diagram shows an edge-to-cloud workflow that includes various technology capabilities that are common in distributed architectures:

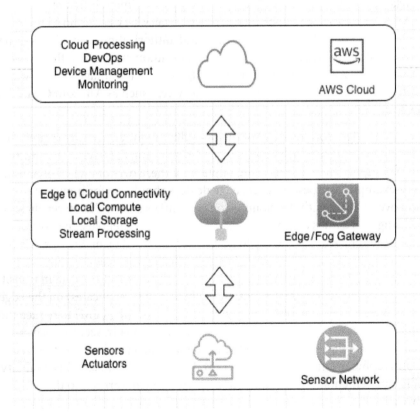

Figure 8.6 – Edge-to-cloud architecture

The edge architecture's standards are still evolving. Considering there are different connectivity interfaces, communication protocols, and topologies, there are heterogeneous ways of solving different use cases. For example, connectivity interfaces may include different short-range (such as *BLE*, *Wi-Fi*, and *Ethernet*) or long-range radio networks (such as *cellular*, *NB-IoT*, and *LoRa*). The connectivity interface that's used needs to be determined during the hardware designing phase and is implemented as a one-time process. Communication protocols may include different transport layer protocols over TCP (connection-oriented such as *MQTT* and *HTTPS*) or UDP (connectionless such as *CoAP*). Recall the layers of the **Open System Interconnection (OSI)** model, which we reviewed in *Chapter 2*, *Foundations of Edge Workloads*. The choice of communication interfaces can be flexible, so long as the underlying Layer 4 protocols are supported on the hardware. For example, if the hardware supports UDP, it can be activated with configuration changes, along with installing additional Layer 7 software (such as a COAP client) as required. Thus, this step can be performed through a cloud-to-edge DevOps workflow (that is, an OTA update). Bringing more intelligence to the edge requires dealing with the challenges of running distributed topologies on a computing infrastructure with low horsepower. Thus, it's necessary to define standards and design principles to design, deploy, and operate optimized software workloads on the edge (such as brokers, microservices, containers, caches, and lightweight databases).

Hopefully, this has helped you understand the unique challenges for edge workloads from a DevOps perspective. In the next section, you will understand how AWS IoT Greengrass can help you build and operate distributed workloads on the edge.

Understanding the DevOps toolchain for the edge

In the previous chapters, you learned how to develop and deploy native processes, data streams, and ML models on the edge locally and then deployed them at scale using **Greengrass's** built-in OTA mechanism. We will explain the reverse approach here; that is, building distributed applications on the cloud using DevOps practices and deploying them to the edge. The following diagram shows the approach to continuously build, test, integrate, and deploy workloads using the **OTA** update mechanism:

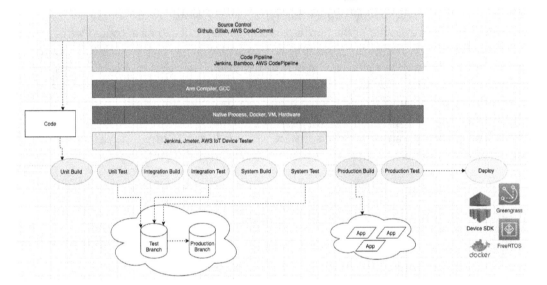

Figure 8.7 – A CI/CD view for Edge applications

The two most common ways to build a distributed architecture on the edge using AWS IoT Greengrass is by using AWS Lambda services or Docker containers.

AWS Lambda at the edge

I want to make it clear, to avoid any confusion, that the concept of Lambda design, which was introduced in *Chapter 5, Ingesting and Streaming Data from the Edge*, is an architectural pattern that's used to operate streaming and batch workflows on the edge. **AWS Lambda**, on the contrary, is a serverless compute service that offers a runtime for executing any type of application with no administration. It allows developers to focus on the business logic, write code in different programming languages (such as *C, C++, Java, Node.js,* and *Go*), and upload it as a ZIP file. The service takes it from there in provisioning the underlying infrastructure's resources and scales based on incoming requests or events.

AWS Lambda has been a popular compute choice in designing event-based architectures for real-time processing, batch, and API-driven workloads. Due to this, AWS has decided to extend the Lambda runtime support for edge processing through **Amazon IoT Greengrass**.

So, are you wondering what the value of implementing AWS Lambda at the edge is?

You are not alone! Considering automated hardware provisioning is not an option for the edge, as explained earlier in this chapter, the value here is around interoperability, consistency, and continuity from the cloud to the edge. It's very common for IoT workloads to have different code bases for the cloud (**distributed stack**) and the edge (**embedded stack**), which leads to additional complexity around code integration, testing, and deployment. This results in additional operational overhead and a delayed time to market.

AWS Lambda aimed to bridge this gap so that the cloud and embedded developers can use similar technology stacks for software development and have interoperable solutions. Therefore, building a DevOps pipeline from the cloud to the edge using a common toolchain becomes feasible.

Benefits of AWS Lambda on AWS IoT Greengrass

There are several benefits of running Lambda functions on the edge, as follows:

- Lambda functions that are deployed locally on the edge devices can *connect to different physical interfaces* such as CANBus, Modbus, or Ethernet to access different serial ports or GPIO on the hardware similar to embedded applications.

- Lambda functions can *act as the glue between different edge components* (such as Stream Manager) within AWS IoT Greengrass and the cloud resources.

- AWS IoT Greengrass also *makes it easier to deploy different versions of Lambda functions* by using an alias or a specific version for the edge. This helps in continuous delivery and is useful for scenarios such as blue/green deployments.

- *Granular access control*, including specifying configurations (run as root) or permissions (read/write) for different local resources (such as disk volumes, serial ports, or GPIOs), can be managed for Lambda functions.

- Lambda functions can be run in both **containerized** and **non-containerized** modes. Non-containerized mode removes the abstraction layer and allows Lambda to run as a regular process on the OS. This is useful for latency-sensitive applications such as ML inferencing.

- Finally, AWS IoT Greengrass *allows you to manage the hardware resources* (RAM) that can be used by the Lambda function on the edge.

The following diagram shows how an AWS Lambda function that's been deployed on the edge can interact with different components on the physical (such as the filesystem) or abstracted layer (such as stream manager on AWS IoT Greengrass):

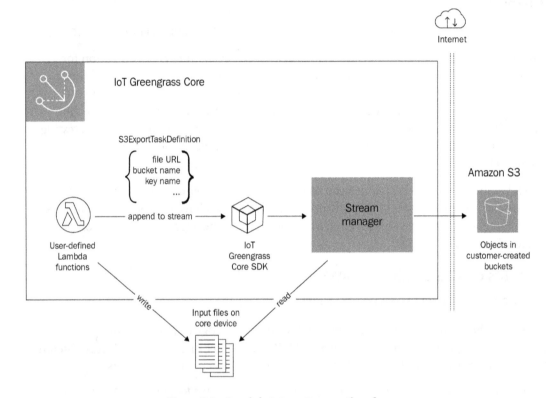

Figure 8.8 – Lambda interactions on the edge

Here, you can see that Lambda provides some distinct value propositions out of the box that you have to build yourself with native processes.

Challenges with Lambda on the edge

As you have understood by now, every solution or architecture has a trade-off. AWS Lambda is not an exception either and can have the following challenges:

- *Lambda functions can be resource-intensive* compared to native processes. This is because they require additional libraries.

- *Lambda functions are AWS only*. Thus, if you are looking to develop a cloud-agnostic edge solution (to mitigate vendor lock-in concerns), you may need to stick to native processes or Docker containers. Although Greengrass v2, as an edge software, is open source, AWS Lambda functions are not.

Now, let's understand containers for the edge.

Containers for the edge

A **container** is a unit of software that packages the necessary code with the required dependencies for the application to run reliably across different computing environments. Essentially, a container provides an abstraction layer to its hosted applications from the underlying OS (such as Ubuntu, Linux, or Windows) or *architecture* (such as x86 or ARM). In addition, since containers are lightweight, a single server or a virtual machine can run multiple containers. For example, you can run a *3-tier architecture* (web, app, and a database) on the same server (or VM) using their respective container images. The two most popular open source frameworks for container management are **Docker** and **Kubernetes**.

In this section, we will primarily discuss Docker as it's the only option that's supported natively by AWS IoT Greengrass at the time of writing. Similar to Lambda, Docker supports an exhaustive set of programming languages and toolchains for the developers to develop, operate, and deploy their applications in an agile fashion. The following diagram shows the reference architecture for a Docker-based workload deployed on AWS IoT Greengrass:

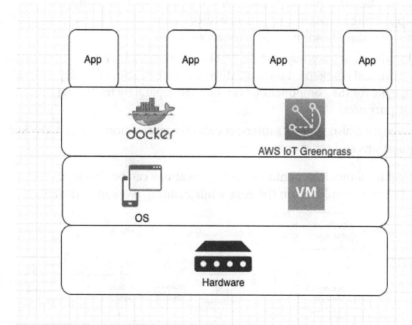

Figure 8.9 – Docker abstraction layers

So, why run containers over Lambda on the edge?

Containers can bring all of the benefits that Lambda does (and more), along with being heterogeneous (different platforms), open source, and better optimized for edge resources. Containers have a broader developer community as well. Since containers have an orchestration and abstraction layer, it's not dependent on other runtimes such as AWS IoT Greengrass. So, if your organization decides to move away to another edge solution, containers are more portable than Lambda functions.

Benefits of Docker containers on AWS IoT Greengrass

Running containers at the edge using Greengrass has the following benefits:

- Developers can continue to use their existing CI/CD pipelines and store artifacts (that is, **Docker images**) in different code repositories such as **Amazon Elastic Container Registry** (**ECR**), the public Docker Hub, the public Docker Trusted Registry, or an S3 bucket.
- Greengrass simplifies deploying to the edge, with the only dependency being having the **Docker image URI** in the configuration. In addition, Greengrass also offers a native Docker application manager component (`aws.greengrass.DockerApplicationManager`) that enables Greengrass to manage credentials and download images from the supported repositories.
- Greengrass offers first-class support for Docker utilities such as `docker-compose`, `docker run`, and `docker load`, all of which can be included as dependencies in the recipe file for the component or can be used separately for testing or monitoring purposes.
- Finally, Greengrass also supports inter-process communication between Docker-based applications and other components.

The following diagram shows how containerized applications can be developed using a CI/CD approach and be deployed on the edge while running AWS IoT Greengrass:

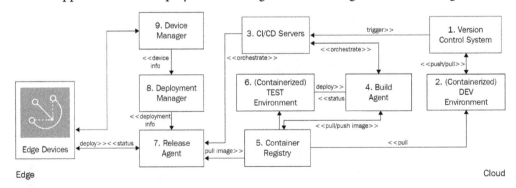

Figure 8.10 – CI/CD approach for Docker workloads

Next, let's learn about the challenges with Docker on the edge.

Challenges with Docker on the edge

Running containers on the edge has some tradeoffs that need to be considered, as follows:

- Managing containers at scale on the edge brings more operational overhead as it can become complex. Thus, it requires careful designing, planning, and monitoring.

- As you build sophisticated edge applications with private and public Docker images, you are increasing the surface area for attacks as well. Thus, adhering to various operational and security best practices at all times is quintessential.

- In addition to AWS IoT Greengrass-specific updates, you need to have a patching and maintenance routine for Docker-specific utilities as well, which, in turn, increases the operational overhead and network charges.

- An additional layer of abstraction with containers may not be a fit for latency-sensitive use cases. For example, performing ML inferencing on GPUs for time-sensitive actions such as detecting an intrusion in your home through computer vision may run better as a native process over a container.

In the lab section of this chapter, you will get your hands dirty by deploying a Docker-based application to the edge using AWS IoT Greengrass.

Additional toolsets for Greengrass deployments

Similar to other AWS services, AWS IoT Greengrass also supports integration with various IaC solutions such as **CloudFormation**, **CDK**, and **Terraform**. All these tools can help you create cloud-based resources and integrate with different CI/CD pipelines for supporting cloud-to-edge deployments.

Now that you are familiar with the benefits and tradeoffs of the DevOps toolchain, let's learn how that extends to machine learning.

MLOps at the edge

Machine Learning Operations (**MLOps**) aims to integrate agile methodologies into the end-to-end process of running machine learning workloads. MLOps brings together best practices from data science, data engineering, and DevOps to streamline model design, development, and delivery across the **machine learning development life cycle (MLDLC)**.

As per MLOps **special interest group** (**SIG**), MLOps is defined as "*The extension of the DevOps methodology to include machine learning and data science assets as first-class citizens within the DevOps ecology.*" MLOps has gained rapid momentum in the last few years from ML practitioners and is a language-, framework-, platform-, and infrastructure-agnostic practice.

The following diagram shows the virtuous cycle of the MLDLC:

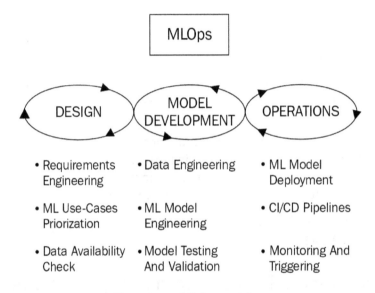

Figure 8.11 – MLOps workflow

The preceding diagram shows how **Operations** is a fundamental block of the ML workflow. We introduced some of the concepts of ML design and development in *Chapter 7, Machine Learning Workloads at the Edge*, so in this section, we will primarily focus on the **Operations** layer.

There are several benefits of MLOps, as follows:

- **Productive**: Data, ML engineers, and data scientists can use self-service environments to iterate faster with curated datasets and integrated ML tools.

- **Repeatable**: Similar to DevOps, bringing automation to all aspects of the ML development life cycle (that is, MLDC) reduces human error and improves efficiency. MLOps helps ensure a repeatable process to help version, build, train, deploy, and operate ML models.

- **Reliable**: Incorporating CI/CD practices into the MLDC adds to the quality and consistency of the deployments.

- **Auditable**: Enabling capabilities such as versioning of all inputs and outputs, ranging from source data to trained models, allows for end-to-end traceability and observability of the ML workload.

- **Governance**: Implementing governance practices to enforce policies helps to guard against model bias and track changes to data lineage and model quality over time.

So, now that you understand what MLOps is, are you curious to know how it's related to IoT and the edge? Let's take a look.

Relevance of MLOps for IoT and the edge

As an IoT/edge SME, you will *NOT* be owning the MLOps process. Rather, you need to ensure that the dependencies are met on the edge (at the hardware and software layer) for the ML engineers to perform their due diligence in setting up and maintaining this workflow. Thus, don't be surprised by the brevity of this section, as our goal is to only introduce you to the fundamental concepts and the associated services available today on AWS for this subject area. We hope to give you a quick ramp-up so that you are adept at having better conversations with ML practitioners in your organization.

So, with that background, let's consider the scenario where the sensors from the connected HBS hub are reporting various anomalies from different customer installations. This is leading to multiple technician calls and thereby impacting the customer experience and bottom line. Thus, your CTO has decided to build a *predictive maintenance solution* using ML models to rapidly identify and fix faults through remote operations. The models should be able to identify data drift dynamically and collect additional information around the reported anomalies. Thus, the goal for ML practitioners here is to build an MLOps workflow so that models can be frequently and automatically trained on the collected data, followed by deploying it to the connected HBS hub.

In addition, it's essential to monitor the performance of the ML models that are deployed on the edge to understand their efficiency; for example, to see how many false positives are being generated. Similar to the DevOps workflow, the ML workflow includes different components such as source control for versioning, a training pipeline for CI/CD, testing for model validation, packaging for deployment, and monitoring for assessing efficiency. If this project is a success, it will help the company add more ML intelligence to the edge and mitigate issues predictively to improve customer experience and reduce costs. The following reference architecture depicts a workflow we can use to implement the predictive maintenance of ML models on AWS IoT Greengrass v2:

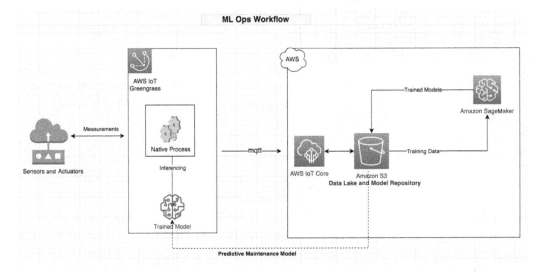

Figure 8.12 – Predictive maintenance of HBS sensors

If we want to implement the preceding architecture, we must try to foresee some common challenges.

MLOps challenges for the edge

Quite often, the most common questions that are asked by edge and ML practitioners related to MLOps are as follows:

- How do I prepare and deploy ML models to edge devices at scale?
- How do I secure the models (being intellectual property) once they've been deployed at the edge?

- How do I monitor the ML models operating at the edge and retrain them when needed?

- How do I eliminate the need for installing resource-intensive runtimes such as TensorFlow and PyTorch?

- How do I interface one or more models with my edge applications using a standard interface?

- How do I automate all these tasks so that there is a repeatable, efficient mechanism in place?

This is not an exhaustive list as it continues to expand with ML being adopted more and more on the edge. The answers to some of those questions are a mix of cultural and technical shifts within an organization. Let's look at some examples:

- **Communication is key**: For MLOps to generate the desired outcomes, communication and collaboration across different stakeholders are key. Considering ML projects involve a different dimension of technology related to algorithms and mathematical models, the ML practitioners often speak a different technical language than traditional IT (or IoT) folks.

  Thus, becoming an ML organization requires time, training, and co-development exercises to be completed across different teams to produce fruitful results.

- **Decoupling and recoupling**: Machine learning models have life cycles that are generally independent of other distributed systems. This decoupling allows ML practitioners to focus on building their applications without being distracted by the rest.

  At the same time, though, ML workflows have certain dependencies, such as on big data workflows or applications required for inferencing. This means that MLOps is a combination of a traditional CI/CD workflow and another workflow engine. This can often become tricky without a robust pipeline and the required toolsets.

- **Deployment can be tricky**: According to Algorithmia's report, *2020 State of Enterprise Machine Learning*, "*Bridging the gap between ML model building and practical deployments is a challenging task.*" There is a fundamental difference between building an ML model in a Jupyter notebook on a laptop or a cloud environment versus deploying that model into a production system that generates value.

With IoT, this problem acts as the force multiplier, as it's required to consider various optimization strategies for different hardware and runtimes before deploying the ML models. For example, in *Chapter 7*, *Machine Learning Workloads at the Edge*, you learned how to optimize ML models using **Amazon SageMaker Neo** so that they can run efficiently in your working environment.

- **The environment matters**: The ML models may need to run in offline conditions and thus are more susceptible to data drift due to the high rate of data from changing environments. For example, think of a scenario where your home has a power or water outage due to a natural disaster. Thus, your devices, such as the HVAC or water pumps, act in an anomalous way, leading to data drift for the locally deployed models. Thus, your locally deployed ML models need to be intelligent enough to handle different false positives in unexpected scenarios.

We have gone through the MLOps challenges for the edge in this section. In the next section, we will understand the MLOps toolchain for the edge.

Understanding the MLOps toolchain for the edge

In *Chapter 7*, *Machine Learning Workloads at the Edge*, you learned how to develop ML models using Amazon SageMaker, optimize them through SageMaker Neo, and deploy them on the edge using AWS IoT Greengrass v2. In this chapter, I would like to introduce you to another service in the SageMaker family called **Edge Manager**, which can help address some of the preceding MLOps challenges and which provides the following capabilities out of the box:

- The ability to automate the build-train-deploy workflow from cloud to edge devices and trace the life cycle of each model.

- Automatically optimize ML models for deployment on a wide variety of edge devices that are powered by CPUs, GPUs, and embedded ML accelerators.

- Supports model compilation from different frameworks such as **DarkNet**, **Keras**, **MXNet**, **PyTorch**, **TensorFlow**, **TensorFlow-Lite**, **ONNX**, and **XGBoost**.

- Supports multi-modal hosting of ML models, along with simple API interfaces, for performing common queries such as loading, unloading, and running inferences on the models on a device.

- Supports open source remote procedure protocols (using **gRPC**), which allow you to integrate with existing edge applications through APIs in common programming languages, such as **Android Java, C#/.NET, Go, Java, Python**, and **C**.

- Offers a dashboard to monitor the performance of models running on different devices across the fleet. So, in the scenario explained earlier with a connected HBS hub, if issues related to model drift, model quality, or predictions are identified, these issues can be quickly visualized in a dashboard or reported through configured alerts.

As you can see, Edge Manager brings robust capabilities to manage required capabilities for MLOps out of the box and brings native integrations with different AWS services, such as AWS IoT Greengrass. The following is a reference architecture for Edge Manager integrating with various other services that you were exposed to earlier in this book, such as SageMaker and S3:

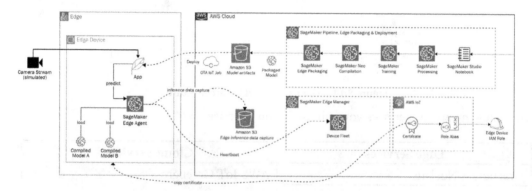

Figure 8.13 – Edge Manager reference architecture

> **Note**
> MLOps is still emerging and can be complicated to implement without the involvement of ML practitioners. If you would like to learn more about this subject, please refer to other books that have been published on this subject.

Now that you have learned the fundamentals of DevOps and MLOps, let's get our hands dirty so that we can apply some of these practices and operate edge workloads in an agile fashion.

Hands-on with the DevOps architecture

In this section, you will learn how to deploy multiple Docker applications to the edge that have already been developed using CI/CD best practices in the cloud. These container images are available in a **Docker repository** called **Docker Hub**. The following diagram shows the architecture for this hands-on exercise, where you will complete *Steps 1* and *2* to integrate the HBS hub with an existing CI/CD pipeline (managed by your DevOps org), configure the Docker containers, and then deploy and validate them so that they can operate at the edge:

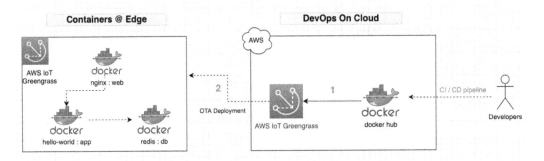

Figure 8.14 – Hands-on DevOps architecture

The following are the services you will use in this exercise:

Edge Services	Cloud Services
Greengrass Nucleus	AWS IoT Greengrass
Docker application	

Figure 8.15 – Services for this exercise

Your objectives for this hands-on section are as follows:

- Deploy container images from Docker Hub to AWS IoT Greengrass.
- Confirm that the containers are running.

Let's get started.

Deploying the container from the cloud to the edge

In this section, you will learn how to deploy a pre-built container from the cloud to the edge:

1. Navigate to your device's terminal and confirm that Docker is installed:

```
cd hbshub/artifacts
docker --version
docker-compose --version
```

If Docker Engine and docker-compose are not installed, please refer to the documentation from Docker for your respective platform to complete this before proceeding.

2. Now, let's review the docker-compose file. If you have not used docker-compose before, then note that it is a utility that's used for defining and running multi-container applications. This tool requires a file called docker-compose.yaml that lists the configuration details for application services to be installed and their dependencies.

3. Please review the docker-compose.yaml file in the artifacts folder of this chapter. It includes three container images from Docker Hub corresponding to the web, application, and database tiers:

```
services:
  web:
    image: "nginx:latest"
  app:
    image: "hello-world:latest"
  db:
    image: "redis:latest"
```

4. Navigate to the following working directory to review the Greengrass recipe file:

```
cd ~/hubsub/recipes
nano com.hbs.hub.Container-1.0.0.yaml
```

5. Note that there is a dependency on the Greengrass-managed Docker application
    manager component. This component helps with retrieving container images from
    the respective repositories and executes Docker-related commands for installing
    and managing the life cycle of containers on the edge:

```
---
RecipeFormatVersion: '2020-01-25'
ComponentName: com.hbs.hub.Container
ComponentVersion: '1.0.0'
ComponentDescription: 'A component that uses Docker
Compose to run images from Docker Hub.'
ComponentPublisher: Amazon
ComponentDependencies:
  aws.greengrass.DockerApplicationManager:
    VersionRequirement: ~2.0.0
Manifests:
  - Platform:
      os: all
    Lifecycle:
      Startup:
        RequiresPrivilege: true
        Script: |
          cd {artifacts:path}
          docker-compose up -d
      Shutdown:
        RequiresPrivilege: true
        Script: |
          cd {artifacts:path}
          docker-compose down
```

6. Now that we have the updated docker-compose file and the Greengrass
    component recipe, let's create a local deployment:

```
sudo /greengrass/v2/bin/greengrass-cli deployment create
--recipeDir ~/hbshub/recipes --artifactDir
~/hbshub/artifacts --merge "com.hbs.hub.Container=1.0.0"
```

7. Verify that the component has been successfully deployed (and is running) using the following command:

```
sudo /greengrass/v2/bin/greengrass-cli component list
```

8. To test which containers are currently running, run the following command:

```
docker container ls
```

You should see the following output:

CONTAINER ID	IMAGE	COMMAND	CREATED	STATUS	PORTS	NAMES
6fa0112921c4	nginx:latest	"/docker-entrypoint..."	2 hours ago	Up 2 hours	80/tcp	100_web_1
6ff39c79ed7d	redis:latest	"docker-entrypoint.s…"	2 hours ago	Up 2 hours	6379/tcp	100_db_1

Figure 8.16 – Running container processes

In our example here, the app (`hello-world`) is a one-time process, so it has already been completed. But the remaining two containers are still up and running. If you want to check all the container processes that have run so far, use the following command:

```
docker ps -a
```

You should see the following output:

CONTAINER ID	IMAGE	COMMAND	CREATED	STATUS	PORTS	NAMES
6fa0112921c4	nginx:latest	"/docker-entrypoint..."	2 hours ago	Up 2 hours	80/tcp	100_web_1
75e849c14379	hello-world:latest	"/hello"	2 hours ago	Exited (0) 2 hours ago		100_app_1
6ff39c79ed7d	redis:latest	"docker-entrypoint.s…"	2 hours ago	Up 2 hours	6379/tcp	100_db_1

Figure 8.17 – All container processes

Congratulations – you now have multiple containers successfully deployed on the edge from a Docker repository (Docker Hub). In the real world, if you want to run a local web application on the HBS hub, this pattern can be useful.

> **Challenge zone (Optional)**
>
> Can you figure out how to deploy a Docker image from Amazon ECR or Amazon S3? Although Docker Hub is useful for storing public container images, enterprises will often use a private repository for their home-grown applications.
>
> Hint: You need to make changes to `docker-compose` with the appropriate URI for the container images and provide the required permissions to the Greengrass role.

With that, you've learned how to deploy any number of containers on the edge (so long as the hardware resource permits it) from heterogeneous sources to develop a multi-faceted architecture on the edge. Let's wrap up this chapter with a quick summary and a set of knowledge check questions.

Summary

In this chapter, you were introduced to the DevOps and MLOps concepts that are required to bring operational efficiency and agility to IoT and ML workloads at the edge. You also learned how to deploy containerized applications from the cloud to the edge. This functionality allowed you to build an intelligent, distributed, and heterogeneous architecture on the Greengrass-enabled HBS hub. With this foundation, your organization can continue to innovate with different kinds of workloads, as well as deliver features and functionalities to the end consumers throughout the life cycle of the product. In the next chapter, you will learn about the best practices of scaling IoT operations as your customer base grows from thousands to millions of devices globally. Specifically, you will learn about the different techniques surrounding fleet provisioning and fleet management that are supported by AWS IoT Greengrass.

Knowledge check

Before moving on to the next chapter, test your knowledge by answering these questions. The answers can be found at the end of the book:

1. What strategy would you implement to bring agility in developing IoT workloads faster?

2. True or false: DevOps is a set of tools to help developers and operations collaborate faster.

3. Can you recall at least two challenges of implementing DevOps with IoT workloads?

4. What services should you consider when designing a DevOps workflow from the cloud to the edge?

5. True or false: Running native containers and AWS Lambda functions on the edge both offer similar benefits.

6. Can you think of at least three benefits of using MLOps with IoT workloads?

7. What are the different phases of an MLOps workflow?

8. True or false: The MLOps toolchain for the edge is limited to a few frameworks and programming languages.

9. What service is available from AWS for performing MLOps on the edge?

References

For more information regarding the topics that were covered in this chapter, take a look at the following resources:

- DevOps and AWS: `https://aws.amazon.com/devops/`

- Infrastructure as Code with AWS CloudFormation: `https://aws.amazon.com/cloudformation/`

- Developing an IoT-MLOps workflow on AWS Using Edge Manager: `https://docs.aws.amazon.com/sagemaker/latest/dg/edge-greengrass.html`

- CRISP-ML Methodology with Quality Assurance: `https://arxiv.org/pdf/2003.05155.pdf`

- Machine Learning Operations: `https://ml-ops.org/`

- Overview of Docker: `https://docs.docker.com/get-started/overview/`

- Different Ways of Running Dockerized Applications on AWS IoT Greengrass: `https://docs.aws.amazon.com/greengrass/v2/developerguide/run-docker-container.html`

- Special interest group: `https://github.com/cdfoundation/sig-mlops`

9
Fleet Management at Scale

The **Internet of things (IoT)** had a humble beginning in 1999, in Procter & Gamble, when Kevin Ashton introduced the idea of integrating a **radio-frequency identification (RFID)** antenna into lipstick shelves to enable branch managers to better track cosmetic inventories for replenishments. Since then, this technology has been adopted across all industry segments in some form or another and has become ubiquitous in today's world.

Managing a set of RFID tags, sensors, and actuators inside a known physical boundary is a relatively easy task. However, managing millions (or billions or trillions) of these devices globally throughout their lifecycle is not. Especially when these devices are spread across different locations with various forms of connectivity and interfaces.

Therefore, in this chapter, you will learn about the best practices of onboarding, maintaining, and diagnosing a fleet of devices remotely through AWS native services. Additionally, you will gain hands-on experience in building an operational hub to assess the health of the connected fleet for taking required actions.

In this chapter, we will be covering the following topics:

- Onboarding a fleet of devices globally

- Managing your fleet at scale

- A hands-on exercise building an operational hub

- Checking your knowledge

Technical requirements

The technical requirements for this chapter are the same as those outlined in *Chapter 2, Foundations of Edge Workloads*. See the full requirements in that chapter.

Onboarding a fleet of devices globally

We already introduced you to the different activities involved in the IoT manufacturing supply chain in *Chapter 8, DevOps and MLOps for the Edge*. **Onboarding** refers to the process of manufacturing, assembling, and registering a device with a registration authority. In this section, we will dive deeper into the following activities that play a part in the onboarding workflow:

Edge	Cloud
• Device assembly • Device bootstrap	• Certificate authority • Device registration • Device activation

Figure 9.1 – Device onboarding activities

So far, in this book, you have been using a **Raspberry Pi** (or a virtual environment) to perform the hands-on exercises. This is a common practice for development and prototyping needs. However, as your project progresses toward a higher environment (such as QA or production), it is recommended that you consider hardware that's industry-grade and can operate in various conditions. Therefore, all the aforementioned activities in *Figure 9.1* need to be completed before your device (that is, the connected HBS hub) can be made available through your distribution channels at different retail stores (and sell like hotcakes!).

For the remainder of this chapter, we assume that your company already has a defined supply chain with your preferred vendors and the device manufacturing workflow is operational to assemble the devices. As your customers unbox these devices (with so much excitement!) and kick off the setup process, the device needs to bootstrap locally (on the edge) and register with the **AWS IoT** services successfully to become fully operational.

So, you must be thinking, what are the necessary steps to perform in advance for the device registration to be a success? Here it comes.

Registering a certificate authority

There are different types of *cryptographic* credentials such as a user ID, password, vended tokens (such as **JWT** and **OAuth**), and symmetric or asymmetric keys that can be used by an IoT device. We recommend using asymmetric keys as, at the time of writing, these are considered to be the most secure approach in the industry. In all the previous hands-on exercises, you took advantage of the **asymmetric X.509** keys and certificates generated by the AWS IoT **Certificate Authority (CA)**, that were embedded in the connected HBS hub running Greengrass. A CA is a trusted entity that issues cryptographic credentials such as digital keys and certificates. These credentials are registered on the cloud and embedded onto the devices to enable transport-layer security or TLS-based mutual authentication. Specifically, there are four digital resources associated with a mutual TLS authentication workflow, as follows:

- **X.509 certificate**: This is a certificate that is required to be present both on the device and in the cloud and is presented during the mutual TLS handshake.

- **Private and public keys**: The asymmetric key pair that is generated using an algorithm such as **Rivest-Shamir-Adleman (RSA)** or **Elliptical Curve cryptography (ECC)**. As a best practice, the private key only stays on the device and should be protected to avoid identity compromises.

- **Signer CA**: This is a root or intermediate certificate that has been issued and signed by a trusted CA such as AWS and Verisign. The device will need to send this issuing CA along with the client certificate during the registration process. If the signer CA is not available, it's also possible to send the **Server Name Indication (SNI)** part of the TLS mutual authentication to validate trust.

- **Server certificate**: This is a certificate that's used by the devices to verify, using the certificate chain presented during the TLS handshake, that it's not communicating with an impersonating server.

The following diagram shows the workflow and the location of these digital resources on the device and in the cloud:

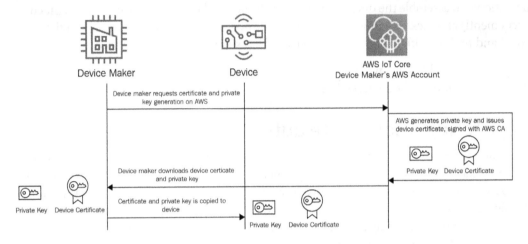

Figure 9.2 – Cryptographic credentials within an IoT workflow

Therefore, it's critical to decide on a CA early on in the design process. This is so that it can issue the aforementioned digital resources required by the devices to perform registration with a backend authority and become fully operational. There are three different ways a CA can be used with AWS IoT Core, as shown in the following table along with a list of pros and cons:

	AWS IoT CA	Bring Your Own CA (BYOCA)	Third-party CA
Cost	Free	Existing or new expense	Existing or new expense
Services	AWS IoT Identity service for IoT Core, Greengrass, and FreeRTOS	Supported AWS services (such as AWS IoT Identity), other enterprise or cloud services	Supported AWS services (such as IoT Identity), other enterprise or cloud services
Scope	Single region only	Multi-region	Multi-region
Operations	Managed service	Overhead of managing the public key infrastructure (aka PKI)	Managed service (Such as AWS Certificate Manager Private Certificate Authority)

Figure 9.3 – Choosing a CA

Once the CA setup is complete, the next step is to choose the device provisioning approach based on the scenario. Let's understand that in more detail next.

Deciding the provisioning approach

Terms such as provisioning and registration are used interchangeably in many contexts in the IoT world, but we believe there is a clear distinction between them. For us, device provisioning is the amalgamation of two activities – device registration and device activation. *Device registration* is the process where a device successfully authenticates using its initial cryptographic credentials against a registration authority (such as the AWS IoT Identity service), reports distinctive attributes such as the model and serial number, and gets associated as a unique identity in a device registry. Additionally, the authority can return a new set of credentials, which the device can replace with the prior ones. Following this, a privilege escalator can enhance the privilege of the associated principal (such as the X.509 certificate) for the device to be activated and fully operational.

There are different approaches to these provisioning steps, which are often derived from the level of control or convenience an organization intends to have in the manufacturing supply chain. Often, this choice is determined by several factors such as in-house skills, cost, security, time to market, or sensitivity to the intellectual property of the product. You learned about the approach of automatic provisioning through IoT Device Tester in *Chapter 2*, *Foundations of Edge Workloads*, which is prevalent for prototyping and experimentation purposes.

In this section, we will discuss two production-grade provisioning approaches that can scale from one device to millions of devices (or more) by working backward from the following scenarios.

Scenario 1 – HBS hubs with unique firmware images are provisioned in bulk

In this scenario, you can provision the fleet of HBS hubs in bulk in your supply chain using unique firmware images that include unique cryptographic credentials:

1. As a device maker of the HBS hub, first, you will associate the CA of your choice with AWS IoT Core using a supported API. This CA will manage the chain of trust and create and validate the cryptographic credentials (such as certificates or certificate signing requests) for the entire fleet of devices.

2. Then, you will create a provisioning template, which, essentially, is a configuration file that holds the instruction for the AWS IoT identity service to create a thing (or the virtual representation) for the fleet of HBS hub devices. This template will include different input parameters such as `ThingName`, `SerialNumber`, and **Certificate Signing Requests** (**CSRs**) that will lead to the creation of IoT resources such as the virtual thing, device metadata, certificates, and policies. You can refer to the template shown on the official *Internet of things on AWS* blog, titled *Deploy Fleets Easily with AWS IoT Device Management Services* (`https://aws.amazon.com/blogs/iot/deploy-fleets-easily-with-aws-iot-device-management-services/`). The template can be found in the *Create Provisioning Template* section.

3. The next step is to generate cryptographic credentials, such as the private keys and CSRs, through preferred tool chains (such as **OpenSSL**). Following this, create an input file with a list of devices and cryptographic credentials, which will be fed to the provisioning template. The input file, in its simplest form, might look like this:

```
{"ThingName": "hbshub-one", "SerialNumber": "0001",
"CSR": "*** CSR FILE CONTENT ***"}

{"ThingName": "hbshub-two", "SerialNumber": "0002",
"CSR": "*** CSR FILE CONTENT ***"}

{"ThingName": "hbshub-three", "SerialNumber": "0003",
"CSR": "*** CSR FILE CONTENT ***"}
```

4. Now, the time has arrived to invoke an API to create all these virtual devices (that is, things), in bulk, in the IoT device registry on the cloud along with its respective cryptographic credentials (such as the certificates or CSRs) that are signed by the CA. Once this step is complete, a token exchange role needs to be created, and these things need to be associated with Greengrass-specific constructs such as a Greengrass core, a thing group, and a configuration file (`config.yaml`) that is required for the installation process.

5. Following this, you will inject the credentials generated in the earlier steps, which include the private key from *step 3* and the signed certificate from *step 4*, into your firmware. This updated firmware, often referred to as the golden image, is then shared with your assembly team (either in-house or a contract manufacturer). The assembly team will flash this image onto the device part of the manufacturing workflow:

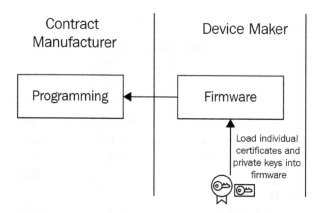

Figure 9.4 – Bulk provisioning with embedded credentials

This approach is pretty common with microcontrollers running an RTOS since incremental updates are not supported (yet) with those hardwares. However, for the connected HBS hub, it's more agile and operationally efficient to decouple the firmware image from the crypto credentials.

That's where this second option comes in. Here, you will still generate the things and unique credentials from your CA in the same way as you did in the previous step, but you will not inject it into the firmware. Instead, you will develop an intelligent firmware that can accept credentials over different interfaces such as **secure shell (SSH)**, a **network file system (NFS)**, or a **serial connection**. As a best practice, it's also common to store the credentials in a separate chip such as a secure element or a **trusted platform module (TPM)**. Additionally, the firmware can use a public key cryptography standards interface (such as *PKCS#11*) to retrieve the keys and certificates as required by the firmware or other local applications in real time. At the time of writing this book, Greengrass v2 is awaiting support for TPM, although it was a supported feature in Greengrass v1:

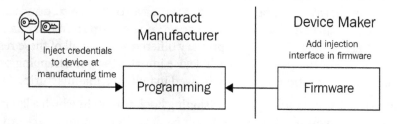

Figure 9.5 – Bulk provisioning with the injection interface

Let us take a look at how to go about the second option:

1. Once the device assembly is complete with the golden image that contains the necessary configurations and credentials, the product reaches the hands of the customer through different distribution channels. When the customer switches on the device and it wakes up for the first time, the bootstrapping process kicks off.

2. The bootstrapping will instantiate various local processes (or *daemons*), that is, the firmware instructions. One of those processes includes registration with AWS IoT Identity services where the device will connect to the cloud endpoint and exchange the embedded credentials part of the TLS mutual authentication. Considering the things, certificates, and policies are already created as a prerequisite to the assembly, the device is fully operational if the mutual authentication is successful.

Scenario 2 – HBS hubs with shared claims are provisioned just in time

Let's consider the following scenario; your devices might not have the capability to accept unique credentials at the time of manufacturing. Or it's cost-prohibitive for your organization to undertake the operational overhead of embedding unique credentials in each HBS hub in your supply chain. This is where another pattern emerges, referred to as fleet provisioning by claim, where, as a device maker, you can embed a non-unique shared credential (referred to as claim) in your fleet. However, we recommend that you do not share the same claim for the entire fleet, rather only a percentage of it to reduce the blast radius in the case of any security issues. Take a look at the following steps:

1. As a first step, the firmware along with a fleet provisioning plugin and a shared certificate (that is, claim) is loaded by the contract manufacturer on the device without performing any customization. The fleet provisioning by claim approach uses a templated methodology that is similar to the previous scenario to provision the required cloud resources. You can refer to a sample template that is provided in AWS's documentation, called *Set up AWS IoT fleet provisioning for Greengrass core devices* (`https://docs.aws.amazon.com/greengrass/v2/developerguide/fleet-provisioning-setup.html#create-provisioning-template`). The primary difference is that all of these resources are provisioned just in time, where each device initiates the bootstrapping process from their current location over being imaged in bulk in a manufacturing facility.

2. This approach also supports a pre-provisioning hook feature in which a lambda function can be invoked to validate different parameters or perform pre-provisioning logic. For example, it can be as simple as overriding a parameter to more complex validations such as checking whether the claim certificate is part of a revocation list:

```python
def pre_provisioning_hook(event, context):
    return {
```

```
'allowProvisioning': True,
'parameterOverrides': {
    'DeviceLocation': 'NewYork'
}
}
```

Here, when the hub wakes up and connects to AWS IoT for the first time, the claim certificate is exchanged for permanent X.509 credentials that have been signed by the CA (AWS or BYO). This is where the fleet provisioning plugin helps, as it allows the device to publish and subscribe to the required **MQ telemetry transport** (**mqtt**) topics, accept the unique credentials, and persist in a secure type of storage:

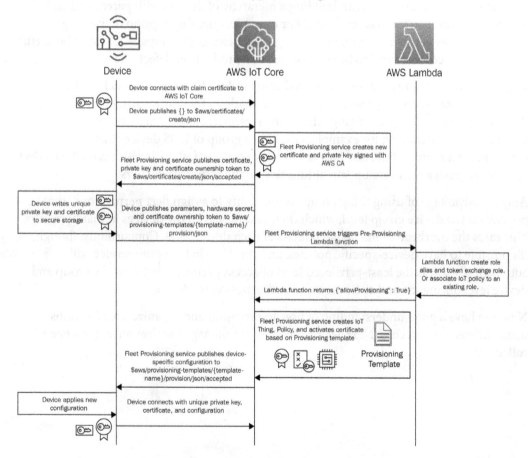

Figure 9.6 – Fleet provisioning with shared claims

3. Following this, the device must also disconnect from the previous session it initiated with the shared claim and reconnect with the unique credentials.

> **Word of Caution**
>
> Fleet provisioning by claim poses security risks if the shared claims are not protected through the supply channels.

Once the devices have been provisioned, the next step is to organize them to ease the management throughout its life cycle. The Greengrass core devices can be organized into thing groups, which is the construct for organizing a fleet of devices within the AWS IoT ecosystem. A **thing group** can be either static or dynamic in nature. As their name suggests, static groups allow the organization of devices based on non-changing attributes such as the product type, the manufacturer, the serial number, the production date, and more.

Additionally, static groups permit building a hierarchy of devices with parent and child devices that can span up to seven layers. For example, querying a group of washing machine sensors within a serial number range that belongs to company XYZ can be useful to identify devices that need to be recalled due to a production defect.

In comparison, dynamic groups are created using indexed information such as the connectivity status, registry metadata, or device shadow. Therefore, the membership of dynamic groups is always changing. That is the reason dynamic groups are not associated with any device hierarchy; for example, querying a group of HBS devices that are connected at a point in time and have a firmware version of v1. This result can allow a fleet operator to push a firmware update notification to the respective owners.

Another advantage of using thing groups is the ability to assign fleet permissions (that is, policies) at the device group level, which then cascades to all the devices in that hierarchy. This eases the overhead of managing policies at each device level. Concurrently, though, it's possible to have device-specific policies, and the AWS IoT Identity service will automatically assess the least-privileged level of access permitted between the group and device level during the authentication and authorization workflow.

Now you have a good understanding of how to provision and organize the HBS hubs using different approaches. Next, let's discuss how to manage the fleet once it has been rolled out.

Managing your device fleet at scale

Although it might be easier to monitor a handful of devices, managing a fleet of devices at scale can turn out to be an operational nightmare. Why? Well, this is because IoT devices (such as the HBS hub) are not just deployed in a controlled perimeter (such as a data center). As you should have gathered by now, these devices can be deployed anywhere, such as home, office, business locations, that might have disparate power utilization, network connectivity, and security postures. For example, there can be times when the devices operate offline and are not available over a public or private network due to the intermittent unavailability of WI-FI connectivity in that premises. Therefore, as an IoT professional, you have to consider various scenarios and plan in advance for managing your fleet at scale.

In the context of a connected HBS hub, device management can help you achieve the following:

- Capture actionable information from the real world.

- Increase efficiency of the solution by capturing anomalies early.

- Optimize cost using predictive or preventive maintenance.

- Prevent the theft of intellectual IP or unauthorized access.

- Build a continuous feedback loop to improve customer experience.

- Generate additional revenue streams with more data insights.

So, as you might have gathered, developing an IoT solution and rolling it out to the customers is just the beginning. It's necessary to govern the entire life cycle of the solution to achieve the business outcomes cited earlier. Therefore, device management can also be considered as a bigger umbrella for the following activities:

- Provision

- Organize

- Monitor

- Maintain

- Diagnose

The following is a diagram showing the IoT Device Management workflow:

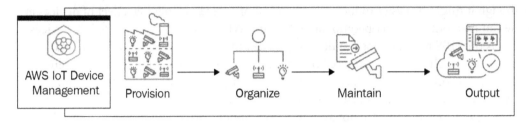

Figure 9.7 – The IoT Device Management workflow

We have already discussed the first three topics in the preceding section in the context of Greengrass. Therefore, we will move on to focus on the remaining activities.

Monitor

Monitoring the Greengrass-enabled HBS hubs and the associated devices will be key in achieving a reliable, highly available, and performant IoT workload. You should have a mechanism to collect monitoring data from the edge solution to debug failures when they occur. Greengrass supports the collection of *system health telemetry* and *custom metrics*, which are diagnostic data points to monitor the performance of critical operations of different components and applications on the Greengrass core devices. The following is a list of the different ways to gather this data:

- **Nucleus emitter** is a Greengrass component that can be deployed to your Greengrass core device to publish telemetry data to a local topic by default:

  - Once the data has been published to a local topic (such as `$local/greengrass/telemetry`), you can act locally on that data on your core device, even when there is intermittent connectivity to the cloud.

  - Optionally, the component can be configured to publish telemetry data to the *mqtt* topic in the cloud.

  - This is a continuous stream of system telemetry data published to a local topic in near real time. The default configuration is every 60 seconds.

  - There are no costs if the telemetry data is published locally, but charges apply when it is pushed to the cloud.

- A **telemetry agent** is another option that you can use to collect local telemetry data, which is enabled by default for all Greengrass core devices:

  - This agent collects telemetry data and publishes it to the cloud on a best effort basis through **Amazon EventBridge**, which is a serverless event bus service.

- Data starts to flow as soon as the Greengrass core device is up and running. By default, the telemetry agent aggregates telemetry data every hour and publishes it to the cloud every 24 hours.

- There are no data transfer charges since the messages are published to reserved topics on AWS IoT Core.

- A *telemetry agent* publishes the following metrics natively:

- System memory and CPU utilization

- The total number of file descriptors, where each descriptor represents an open file

- The number of components in various states, such as the following:

  - Installed, new, and starting

  - Running, stopping, and finishing

  - Errored, broken, and stateless

Later, in the hands-on section, you will collect these metrics and process them on the cloud.

- *CloudWatch metrics* is a Greengrass component that allows you to publish custom metrics to Amazon CloudWatch:

- Any custom component such as lambda functions or containers deployed on the core device can publish the custom metrics to a local topic (`cloudwatch/metric/put`).

- The components can be configured to specify different publish intervals (in seconds), so the metrics can be published with or without batching. For example, with lambda, the default batching window is 10 seconds and the maximum wait time could be 900 seconds.

- So, if you think of scenarios where you have to collect metrics from the sensors and actuators associated with the hub and not just from the gateways, a custom application can retrieve those data points and publish them locally or to cloud endpoints for monitoring purposes.

- *Log manager* is a Greengrass component that can be deployed to your Greengrass core device to collect and, optionally, upload logs to Amazon CloudWatch Logs:

- Although metrics can help reflect the state of the device, logs are critical for troubleshooting exceptions or failures.

- For the real-time observability of logs, Greengrass offers various log files. Some of these you might have already used by now, such as the following:

  - **Greengrass.log**: This is a log file that is used to view real-time information about nucleus and component deployments. For example, with an HBS hub, this log file can report the errors, exceptions, and failures of the nucleus software, which you (or the customer) can analyze for downtimes or malfunctions.

  - **Component.log**: This is a log file(s) to view real-time information about the components running on the core device.

  - **Main.log**: This is a log file that handles the component life cycle information.

- The log manager component can upload logs in various frequencies. The default configuration for log manager is to upload new logs every 5 minutes to AWS CloudWatch. Additionally, it's possible to configure a lower upload interval for more frequent uploads. The log format is also configurable between text format and JSON format.

- Log manager also supports file rotation every hour or when the file size limit has exceeded. The default size limit for the log files is 1 MB and the disk size is 10 MB and is fully configurable. To optimize log sizes, it's also a best practice to use different verbosity levels for different environments (such as development, testing, and production):

  - For example, you can choose DEBUG-level logs to help with troubleshooting in non-production environments or ERROR-level logs to reduce the number of logs that a device generates in a production environment. This choice also helps to optimize costs.

As you are collecting all of these data points (metrics and logs) from the HBS hub and publishing them to the cloud, the next step is to allow different personas such as fleet operators (or other downstream businesses) to consume this information. This can be achieved through CloudWatch, which natively offers various capabilities related to logging insights, generating dashboards, setting up alarms, and more. If your organization has already standardized on a monitoring solution (such as **Splunk**, **Sumologic**, **Datadog**, or others) CloudWatch also supports that integration.

Finally, in the control plane, Greengrass integrates with **AWS CloudTrail** to log access events related to service APIs, IAM users, and roles. These access logs can be used to determine additional details about Greengrass access such as the IP address from which a request was made, who the request was made by, and when it was made, which can be useful for various security and operational needs.

Maintenance

The previously explained services, such as Amazon CloudWatch (or a third-party solution), can be robust enough to generate the various insights required to monitor the health of IoT workloads. However, another common ask from IoT administrators or fleet operators is to have a single-pane-of-glass view that allows them to consume a comprehensive set of information from the device fleet, to quickly troubleshoot operational events.

For example, consider a scenario where customers are complaining that their HBS hubs are malfunctioning. As a fleet operator, you can observe a lot of connection drops and high-resource utilization from the dashboard. Therefore, you look up the logs (on a device or in the cloud) and identify it as a memory leak issue due to a specific component (such as *Aggregator*). Based on your operation playbook, you need to identify whether this is a one-off issue or whether more devices in the fleet are showing similar behavior. Therefore, you need an interface to search, identify, and visualize the metrics such as the device state, device connection, battery level across the fleet, or on a set filtered by user location. Here comes the need for a fleet management solution such as **AWS Fleet Hub**, which allows the creation of a fully managed web application to cater to various personas using a no-code approach. In our scenario, this web application can help the operators to view, query, and interact with a fleet of connected HBS hubs in near real time and troubleshoot the issue further. In addition to monitoring, the operators can also respond to alarms and trigger a remote operation **over the air** (**OTA**) to remediate deployed devices from a single interface. AWS Fleet Hub applications also enable the following:

- *Integration* with existing identity and access management systems such as Active Directory and LDAP, which allow role-based access to different personas such as the fleet operators, fleet managers, device makers, third parties, and IT operators who are interacting with the HBS hub in some way. Additionally, this allows these users to use **single sign-on** (**SSO**) and access a fleet hub from any browser on a desktop, tablet, or smartphone.

- *Aggregation* of data from other services such as AWS IoT Fleet Indexing, Amazon CloudWatch, or **Amazon Simple Notification Service** (**SNS**). The IoT Fleet Indexing service helps to index, search, query, and aggregate data from device registry, device shadow, and device connectivity events. Also, it's possible to create custom fields. CloudWatch metrics can be used in combination with these searchable fields to create alarms. Finally, Amazon SNS can notify different personas when an alarm threshold has been breached.

In summary, these capabilities from Fleet Hub can allow an organization to respond more quickly to different operational events and, thereby, improve customer experience.

Diagnose

In the preceding scenario, you learned how a fleet operator can stay well informed about the operational events in near real time through a single-pane-of-glass view. However, what about diagnosing the issue further if the remote actions through Fleet Hub are not sufficient to remediate the identified issue? For example, an operator might have triggered a remote action to restart the aggregator component or the HBS hub itself, but that did not solve the problem for the end consumer. Therefore, as a next step, the operator is required to gain direct access to the hub or associated sensors for further troubleshooting. Traditionally, in such a situation, a company will schedule an appointment with a technician, which means additional cost and wait time for the customers. That's where a remote diagnostics capability such as **AWS IoT Secure Tunneling** can be useful. This is an AWS Managed service that allows fleet operators to gain additional privileges (such as SSH or RDP access) over a secure tunnel to the destination device.

The secure tunneling component of Greengrass enables secure bidirectional communication between an operator workstation and a Greengrass-enabled device (such as the HBS hub) even if it's behind restricted firewalls. This is made possible because the remote operations navigate through a secure tunnel under the hood. Moreover, the devices will also continue to use the same cryptographic credentials (that is, *X509 certificates*) used in telemetry for this remote operation. The only other dependency from the client side (that is, the **fleet operator**) is the installation of proxy software on the laptop or a web browser. That is because this proxy software makes the magic happen by allowing the exchange of temporary credentials (that is, **access tokens**) with the tunneling service when the sessions are initiated. The following diagram shows the workflow of secure tunneling:

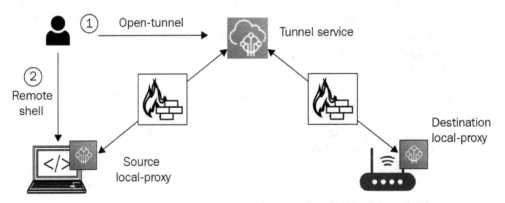

Figure 9.8 – The secure tunneling workflow for diagnostics

For our scenario, the source refers to the workstation of the fleet operator, the destination refers to the connected HBS hub, and the secure tunnel service is managed by AWS.

Now that you have gained a good understanding of how to better monitor, maintain, and diagnose edge devices, let's get our hands dirty in the final section of this chapter.

Getting hands-on with Fleet Hub architecture

In this section, you will learn how to use the nucleus emitter and the telemetry agent to capture various metrics and logs from edge devices and visualize those through Amazon CloudWatch and AWS IoT Fleet Hub. The following is the architecture that shows the different services and steps you will complete during the lab:

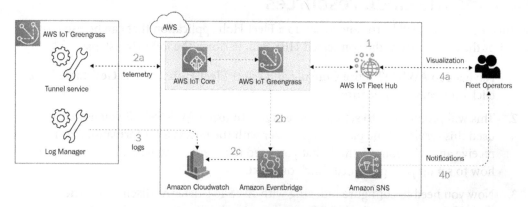

Figure 9.9 – Hands-on operational hub

The following table lists the services that you will use in this exercise:

Edge services	Cloud services
Greengrass Nucleus	AWS IoT Fleet Hub
Greengrass Nucleus Emitter	AWS IoT EventBridge
Greengrass Log Manager	Amazon CloudWatch

Figure 9.10 – The services in scope for this exercise

Your objective in this hands-on section includes the following steps, as depicted in the preceding architecture:

1. Build an operational dashboard using AWS IoT Fleet Hub.

2. Deploy a nucleus emitter component and collect metrics through the telemetry from the edge.

3. Deploy a log manager component and stream the logs to Cloudwatch.

4. Visualize the results on IoT Fleet Hub and CloudWatch.

Let's take a look at the preceding steps, in more detail, next.

Building the cloud resources

In this section, you will learn how to set up a Fleet Hub application that can be used to monitor the metrics from the connected HBS hub. Perform the following steps:

1. Navigate to **AWS IoT Core Console** and select **Fleet Hub**. Choose **Get started** and click on **Create Application**.

2. This will prompt you to set up **Single Sign-On** using AWS SSO. If you have not used this service in the past, create a user with the necessary information. You will receive an invitation via email that you need to accept along with instructions about how to set up your password. Click on **Next**.

3. Now you need to configure indexing with AWS IoT data. As discussed earlier, Fleet Hub gives you a single-pane-of-glass view by aggregating information from different sources. This is where you set all of these integrations.

4. Click on **Manage Indexing** in the **Fleet indexing** section. This will open up the AWS IoT settings page. Click on **Manage Indexing** again and enable all the available options such as **Thing Indexing** and **Thing Group Indexing**. Optionally, you can also create custom search fields if you wish. Click on **Update**.

5. Navigate back to the Fleet Hub setup screen and the settings should be in active status now. Click on **Next**.

6. Create an application role and an application with a name of your choice. Click on the view policy permissions to understand the access provided to the application. You should notice that you are providing access to the IoT, CloudWatch, and SNS resources for Fleet Hub to integrate with all of these sources, as mentioned earlier. Click on **Create application**.

7. On the **Applications** tab of Fleet Hub, click on the application URL once it shows active. For the first access, this will prompt you to add an SSO user. Click on that and add the user you created earlier. Click on **Add selected users**.

8. Once it's complete, click on the application URL again, proceed to the web dashboard, and sign in with the credentials you configured in *step 2*. Click on the application icon, and it should open up the dashboard for you.

Congratulations! You are all set up with the Fleet Hub dashboard!

> **Note**
> Although we have only created one user for this lab, you can integrate AWS SSO with your organization's identity management systems such as Active Directory. This will allow role-based access to the dashboard for different personas. Ideally, this configuration will fall under the purview of identity engineers and won't be the responsibility of the IoT professionals.

Next, let's set up the routing rules for ingesting the telemetry data from the HBS hub to the cloud:

1. Navigate to the **Amazon EventBridge** console (`https://go.aws/3xcB2T7`), and click on **Create rule**.

2. Choose a name and description for the rule.

3. In the **Define pattern** section, configure the following:

   - Check the event pattern and predefined pattern by service.

   - Service provider: `AWS`.

   - Service name: `Greengrass`.

   - Event type: `Greengrass Telemetry Data`.

4. Keep the **Select an Event Bus** section in its default setting.

5. Under **Select targets**, configure the target to be a CloudWatch group:

   - **Target**: CloudWatch log group

   - **Log Group**: `/aws/events/`<replace this with a name of your choice>

6. Keep everything else in its default setting and click on **Create**.

Great work! EventBridge is all set to ingest telemetry data and publish it to the CloudWatch group.

We will revisit these dashboards again at the end of the lab to visualize the collected data from the hubs. For now, let's switch gears to configure and deploy the edge connectors on Greengrass.

Deploying the components from the cloud to the edge

In this section, you will learn how to deploy the **Nucleus Emitter** and **Log Manager** agents to a Greengrass-enabled HBS hub, to publish the health telemetry to the cloud. Perform the following steps:

1. Please navigate to the Amazon Greengrass console and choose **Deployments**. Choose the existing deployment, and click on **Revise**. On the **Select Components** page, in the **Public components** section, search for and select the `aws.greengrass.telemetry.NucleusEmitter` and `aws.greengrass.LogManager` components. Feel free to click on **View recipe** to review the configuration of this component.

2. Click on **Next** by keeping all the previous components along with the preceding two components selected. Keep the other options in their default settings in the following screens and choose **Deploy**.

Visualizing the results

Now you have set up the fleet hub and deployed the agents on the edge, you can visualize the health telemetry data using the following steps:

1. Navigate to **AWS IoT Core Console**, click on **Fleet Hub**, and select the application you created earlier in the lab.

2. In the **Device list** section, click on the Greengrass-enabled HBS hub device. You will be able to view the status of the device along with various other attributes such as the field attributes, the device shadow file (that is, the device state if configured), the group to which the device belongs, and the history of the deployments (that is, the jobs).

3. Navigate back to the home screen. Feel free to play with the search and filter options provided at the top of the screen to refine your results. Generally, this is useful when you have a large number of gateways and associated devices. You can also flip to the **Summary** section to visualize the total number of devices by thing types, the thing groups, and the reasons to disconnect.

4. Click on **Create alarm**. This will help you set up notifications for the following breach:

   I.   Choose a field: **Disconnect Reason**.

   II.   Choose an aggregation type: **Count**.

   III.   Choose a **Period** setting of **5 minutes**, then click on **Next**.

   IV.   Choose the **Metric** setting to be **Greater/Equal than 1**, then click on **Next**.

   V.   Assuming you have only one device gateway, you can increase the count if you have more.

   VI.   Configure the notify and alarm details, then click on **Next and Submit**.

   Therefore, as a fleet operator, you can now visualize the health of your device fleet along with being alarmed for threshold breaches.

5. Navigate to the **AWS CloudWatch** console, click on **Logs** and then select **Log groups**.

6. Search for and click on the /aws/events/<replace this with a name of your choice> log group, and visualize the log streams.

7. It might take some time to populate, but the log streams should show the telemetry data collected from the Greengrass hub device.

8. Feel free to play with **Log Insights**, which allows you to analyze the logs through a query interface.

So far, you have learned how to operate a fleet of Greengrass-enabled devices using AWS native solutions. These patterns are also applicable for non-Greengrass devices, for example, the devices that leverage other devices' SDKs (such as *AWS IoT Device SDK* or *FreeRTOS*).

> **Challenge Zone**
>
> I would like to throw a quick challenge for you to determine how you can trigger an OTA job from AWS IoT Fleet Hub to a specific HBS hub device. This can be useful when you have to push an update such as a configuration file that is required during an operational event. Best of luck!

Let's wrap up this chapter with a quick summary and a set of knowledge-check questions.

Summary

In this chapter, you were introduced to design patterns and the best practices of onboarding, managing, and maintaining a fleet of devices. These practices can help you to provision and operate millions (or more) of connected devices across different geographic locations. Additionally, you learned how to diagnose edge devices remotely for common problems or tunnel in securely for advanced troubleshooting. Following this, you implemented an edge-to-cloud architecture, leveraging various AWS-built components and services. This allowed you to collect health telemetry from the HBS hubs, which a fleet operator can visualize through dashboards, be notified through alarms, or take action as required.

In the next and final chapter, we will summarize all the key lessons that you have learned throughout the book (and more), so you are all set to build well-architected solutions for the real world.

Knowledge check

Before moving on to the next chapter, test your knowledge by answering these questions. The answers can be found at the end of the book:

1. True or false: Device registration and device activation are the same.

2. What are the different ways to leverage a CA with AWS IoT Greengrass?

3. Is there an option to provision devices in real time? If yes, then what is it?

4. True or false: Metrics and logs are the only data points required to monitor an IoT workload.

5. What do you think is a benefit of having a single-pane-of-glass view for your entire fleet of devices?

6. What is a mitigation strategy for remote troubleshooting devices without sending technicians if required? (Hint: think tunnel.)

7. What components does AWS IoT Greengrass provide to collect system health telemetry?

8. True or false: Aggregation of metrics on the edge device is not possible. It can only be done in the cloud.

References

Take a look at the following resources for additional information on the concepts discussed in this chapter:

- Whitepaper on *Device Manufacturing and Provisioning with X.509 Certificates in AWS IoT Core*: `https://d1.awsstatic.com/whitepapers/device-manufacturing-provisioning.pdf`

- When to use AWS IoT device management: `https://aws.amazon.com/iot-device-management/`

- The AWS IoT Greengrass launch of fleet management capabilities: `https://aws.amazon.com/about-aws/whats-new/2021/08/aws-iot-greengrass-v-2-4/`

- Fleet Hub for AWS IoT Device Management: `https://docs.aws.amazon.com/iot/latest/fleethubuserguide/what-is-aws-iot-monitor.html`

- AWS IoT thing groups: `https://docs.aws.amazon.com/iot/latest/developerguide/thing-groups.html`

Section 4: Bring It All Together

This section will conclude the hands-on project with a review of what makes a solution well architected (security, reliability, operational excellence, cost optimization, and performance) and guidance on how to use what you have learned from this book to deliver your next outcome. It will guide you on what to do next.

This section comprises the following chapter:

- *Chapter 10, Reviewing the Solution with AWS Well-Architected Framework*

10
Reviewing the Solution with AWS Well-Architected Framework

You now have the skills required to create edge **machine learning** (ML) solutions. This chapter acts as both a summary of the key lessons that have been learned throughout this book and follows through on why they are best practices by reviewing the delivered solution. By reviewing the solution, we can see how the Home Base Solutions prototype hub design holds up and where there are further opportunities to improve it. You will learn what it is like to perform a deep analysis of the solution using the **AWS Well-Architected Framework**, a mechanism that was created for reviewing complex solutions. Finally, we'll leave you with suggested next steps for your journey as a practitioner of delivering intelligent workloads to the edge.

In this chapter, we're going to cover the following main topics:

- Summarizing the key lessons

- Describing the AWS Well-Architected Framework

- Reviewing the solution

- Diving deeper into AWS services

Summarizing the key lessons

In this section, we will group and summarize the key lessons from throughout this book as a quick reference to ensure that the most important lessons were not missed. There is a loose chronology to the groupings based on the material from *Chapters 1 to 9*, but some lessons may appear in a group outside the order in which they appeared in this book.

Defining edge ML solutions

The following key lessons capture the definition, value proposition, and shape of an edge ML solution:

- **Definition of an edge ML solution**: Bringing intelligent workloads to the edge means applying ML technology that's been incorporated into cyber-physical solutions that interoperate the analog and digital spaces. An edge ML solution uses devices that have sufficient compute power to run ML workloads and either directly interface with physical components such as sensors and actuators, or indirectly interface with end devices over a local network or serial protocol.

- **Key benefits for ML at the edge**: The four key benefits of bringing intelligent workloads to the edge are reducing the latency of reacting to a measured event, improving the solution's availability by reducing its runtime dependency on a remote network entity such as a server, reducing the cost of operations by reducing the quantity of data to transmit over the WAN, and enabling compliance with data governance policies by handling protected data exclusively at the edge.

- **Tools for edge ML solutions**: The three tools that are needed to operate intelligent workloads at the edge are a runtime for orchestrating your edge software, a ML library and ML model, and a mechanism for exchanging data bi-directionally throughout the edge and with remote services.

- **Decoupled, isolated services**: Design your edge ML solutions using principles from the service-oriented architecture to deliver a cohesive whole made up of isolated services that use decoupling mechanisms to interact. Code that's been designed with a singular purpose is easier to write, test, reuse, and maintain. The code that acquires measurements from a sensor does not need to know how to directly invoke a ML inference service. The inference service does not need to know how to emit results directly to a connected actuator. The degree of isolation and separation of concerns to achieve is a spectrum and a balancing act for architects to consider trade-offs.

- **Don't re-engineer solved problems**: Use proven, trusted technologies to solve implementation details that aren't core to the business problems that are solved by your edge ML solution. For example, don't create a new messaging protocol or data storage layer unless none already satisfy your business requirements.

- **Common edge topologies**: Four topologies that are common in building edge solutions are **star**, **bus**, **tree**, and **hybrid**. The star topology describes how leaf devices interact with a hub or gateway device that tends to run any ML workloads. The bus topology describes how isolated services can interact with each other using decoupling mechanisms. The tree topology describes how a fleet of edge solutions can be managed from a central service. The hybrid topology describes the general shape of any macro-level architecture of edge solutions interacting with cloud services.

Using IoT Greengrass

The following key lessons summarize the definition of **AWS IoT Greengrass** and the best practices for using it to deliver edge ML solutions:

- **What is Greengrass?** AWS IoT Greengrass is a runtime orchestration tool designed for delivering intelligent workloads to the edge by solving many of the implementation details common to IoT and ML solutions, enabling architects to focus on rapidly delivering business objectives. Greengrass supports a service-oriented architecture by letting developers define components that run independently and optionally interact with the wider solution by using decoupled mechanisms such as interprocess communication channels, streams, files, or message queues. Greengrass natively interacts with AWS services to deliver common functionality that architects would otherwise need to solve for, such as deploying software, fetching resources, and transmitting data to the cloud.

- **Building with components**: Greengrass defines units of functionality as **components**, which are recipes for bundling static resources such as artifacts, configuration, dependencies, and runtime behavior. Developers can run any kind of code as a component, be it a shell script, interpreted code, a compiled binary, an AWS Lambda function, or a container such as Docker.

- **Deploying components to the edge**: Components can be deployed locally during development cycles using the Greengrass CLI on a test device. For production use, components are registered in the AWS IoT Greengrass service and deployed remotely as part of a rollout to one or more grouped devices. A **deployment** defines a set of components to include the version of the component in and optionally any configuration to apply to those components at the time of deployment. A device can belong to multiple groups and will aggregate all the current deployments of the groups that it belongs to.

- **Security model between the edge and the cloud**: The security model between a Greengrass device and AWS services uses a combination of asymmetric cryptography, roles, and policies. Greengrass identifies itself to AWS IoT services using a private key and certificate registered with AWS IoT Core. This certificate is attached to an IoT policy that grants permissions, such as connecting and exchanging messages. The device can request temporary AWS credentials from the **AWS IoT Core** credentials provider service to identify itself with other AWS services. This works by specifying an **AWS Identity and Access Management** role that has policies attached to it to grant permissions to other AWS services. Before you add another AWS service interaction to your Greengrass solution, you need to attach a new policy or update an attached policy to include the appropriate permission for that API.

- **Accelerate building with managed components**: Use AWS-managed components to solve requirements when applicable. These components solve common requirements such as interacting with AWS services, deploying a local MQTT broker to connect to local devices, synchronizing device state between the edge and the cloud, and running ML workloads.

Modeling data and ML workloads

The following key lessons summarize the techniques and patterns you should consider when breaking down a problem into modeled data and the ML workloads that you use in your edge ML solutions:

- **Types of structured data**: Data that's acquired at the edge can be classified into three types: **structured** (a well-defined schema), **semi-structured** (a schema with some variance in terms of used keys), and **unstructured data** (a schema with high variance or no schema). All three kinds of data can be evaluated by ML workloads, but the training methods and algorithms may differ for each.

- **Analyze data to select implementation choices**: Use data modeling techniques to break down a high-level problem from the conceptual model, to a logical model, to a physical model to inform implementation decisions when choosing technologies for collecting, storing, and accessing data. Analyze your data's size, shape, velocity, and consistency requirements to inform implementation decisions when choosing data storage technologies.

- **Common data flow patterns**: Some of the common data flow patterns that can be used in an edge ML architecture are **extract, transform, load (ETL)**, **event-driven (streaming)**, **micro-batch processing**, and **Lambda architecture (parallel hot/cold paths)**. Avoid anti-patterns for edge architecture such as complex event detection, batch processing, data replication, and data archiving. These patterns are best implemented at layers in your topologies, such as a data center or cloud services.

- **Domain-driven design**: Consider the 10 principles of domain-driven design to best organize your data: manage data ownership through domains, define domains using bounded contexts, link a bounded context to application workloads, share the ubiquitous language within the bounded context, preserve the original sourced data, associate the data with metadata, use the right tool for the right job, tier your data storage, secure and govern the data pipeline, and design for scale.

- **The three laws of edge workloads**: Keep data workloads at the edge (instead of the cloud) when you must observe the three laws. The **law of physics** means that the latency between the edge and the cloud has limits, and sometimes your workload requirements cannot tolerate this latency. The **law of economics** means it may be cost-prohibitive to move all your data to the cloud. The **law of the land** means that there are data governance and compliance requirements that necessitate that some data remains at the edge.

- **Types of ML training algorithms**: ML models can be trained with one of three patterns: **supervised** (the training data is labeled by a human), **unsupervised** (the training data is unlabelled; the machine finds patterns or conclusions on its own), or **semi-supervised** (a mix of labeled and unlabelled data). Training a model to mimic the work of a human expert, such as classifying objects in an image, typically means using a supervised or semi-supervised pattern. Training a model to find relationships between data typically means using an unsupervised pattern.

- **Iterating the data-to-model life cycle**: Use the **Cross-Industry Standard Process for Data Mining (CRISP-DM)** to iterate your ML workloads, from understanding your data to preparing it for training, to evaluating model performance, and then deploying models to the edge.

- **Use ML appropriately**: Not every problem can or should be solved with ML. Small datasets or data with low signal-to-noise ratios tend not to train useful models. Simple requirements (such as needing a one-off prediction) can be solved with conventional data analysis, querying, and regression techniques.

- **Value of the cloud in training models**: Use the scale of the cloud to train models efficiently and on sufficiently large datasets. Once your models are performing well in the evaluation phase, use model optimization to compress the model so that it has high efficiency and a small footprint on the target hardware platform running your edge ML solution. Continue to test and evaluate the performance of your compressed model on the device and after any retraining events.

- **ML needs a team**: A single technical resource can push all the buttons needed to gather data, train a model, and deploy it to the edge, but the process of training an effective model is multi-faceted. Training effective models and deploying them to the edge requires experts from multiple domains to reach a successful outcome. It's okay that one person can't do it all.

Operating a production solution

The following key lessons summarize important distinctions in the production version of your solution and how to operate the solution at scale:

- **DevOps is cultural**: **Developer operations** (**DevOps**) is not just about new technology and tools. It represents a cultural shift in how organizations promote ownership, collaboration, and cohesiveness across teams to foster innovation. The DevOps paradigm yields benefits to the software delivery life cycle of edge ML solutions, in addition to traditional software development.

- **Use managed components for monitoring**: Use components provided by AWS to store logs and metrics in your **Amazon CloudWatch** account. This will help your team operate a fleet of devices by diagnosing issues remotely through the logs and monitoring for unhealthy devices with alarms on the metrics.

- **IaC is valuable for the edge, too**: Store and deploy your solution as **Infrastructure as Code (IaC)** resources where possible. This makes it easier to maintain your solution's definition and reliably reproduce results across deployments.

- **Your device life cycle begins with manufacturing**: Providing identities to devices and defining their provisioning processes has implications for your device's supply chain. Provisioning a test device on your desk is easy. Creating a provisioning pipeline for a production fleet is much more challenging. Communicate the requirements early with your **supply chain vendors**, **original equipment manufacturers (OEMs)**, and **original device manufacturers (ODMs)**.

- **Ship code in virtualized environments**: Your software components can be defined as scripts, source code, binaries, and anything in between. Prefer to ship your code in virtualized environments such as Docker and AWS Lambda where possible to deliver more predictability for runtime operations at the edge.

- **MLOps is circular**: Much like the CRISP-DM model for solving problems with data science, the pattern of building and operating ML models is circular. MLOps with models deployed to the edge can be extra challenging as devices are often remote, offline, or exposed to unpredictable elements. Design MLOps into your product life cycle early to lean into good habits. Adding it later is only harder.

- **Deployments can be expensive**: Services such as AWS IoT Greengrass make it easy to deploy software to the edge, but the cost of transmitting data must be considered. Many edge solutions are at the end of expensive network connections where you cannot afford to incrementally push models over and over again or fix broken deployments. Set up your DevOps and MLOps pipelines so that you have the highest confidence in your deployments before they go out to the production fleet.

- **Scale the provisioning process**: **Certificate authorities (CAs)** let you define your device's identity with unique certificates. Use your own CA, one from a trusted vendor, or the one provided by AWS to scale up the identities of your device fleets. Use automated provisioning strategies such as **Just-in-Time (JIT)** provisioning to onboard devices as they connect to your service for the first time.

- **Operators need to scale, too**: Scaled production fleets of devices can mean managing thousands to millions of devices. Use tooling that simplifies how to operate that many entities by focusing on outliers and high severity issues. This means you need a solution that captures and indexes this kind of operational data. You also need a solution that makes it easy to dive deep into a single device or apply a fix to a large selection of impacted devices at a time.

In the next section, you will learn about a framework provided by AWS for evaluating design trade-offs when building solutions on the platform.

Describing the AWS Well-Architected Framework

In 2015, **AWS** launched a framework for guiding developers through the process of making good design decisions when building on AWS. The AWS Well-Architected Framework codifies the best practices for defining, deploying, and operating workloads on the AWS cloud. It exists as a whitepaper of best practices and a web-based tool to approach a solution evaluation as a checklist of considerations and suggested mitigation strategies. This expertise aims to serve AWS customers but is delivered in a format that is generally useful for evaluating any kind of digital workload. We will use this framework to retroactively review this book's solution of the Home Base Solutions appliance monitoring product.

The Well-Architected Framework organizes best practices into five pillars. A **pillar** is a section of the framework that aggregates design principles and guiding questions to resolve under a common purpose. The five pillars are as follows:

- **Operational excellence**
- **Security**
- **Reliability**
- **Performance efficiency**
- **Cost optimization**

You may recognize some of these pillars as the key benefits we used to define the value proposition of edge ML solutions in *Chapter 1, Introduction to the Data-Driven Edge with Machine Learning*! Each pillar includes a narrative and a set of questions to evaluate and consider. The questions that the architect does not have a clear response or existing mitigation strategy for are then used to define the gap between how well the solution is architected now and where it needs to be. For example, if the review helps us identify a single point of failure in our architecture, then we would decide between the acceptability of that risk in our solution or whether to refactor with a failover mechanism.

It's important to understand that when you're answering the framework's questions to review your solution, there are no objectively right or wrong answers. The overall posture of your solution is not a quantifiable outcome of completing a review. The process you use to answer individual questions may identify important refactors or highlight gaps in the original design. It's up to your team's *definition of done* to decide how complete or thorough your answers must be and how many questions are resolved, in the sense that your team is satisfied with the due diligence that's been performed. A lazy or superficial review may not lead to any meaningful change. As the criticality of the solution increases, the amount of rigor in your review may scale proportionally or even non-linearly.

In your application of the framework, you may find value in moving pillar by pillar, answering each question in series, or by crafting a subset of prioritized questions as a cross-section of all the pillars. It is also recommended and more common to review the framework between the steps of designing the solution and implementing it. This helps architects prevent failures and raise the security posture before investing time and resources in building the solution. For this book, we elected to save the review for the end to move quickly into the hands-on projects, recognizing that we are practicing in a safe, prototype environment.

The AWS Well-Architected Framework also includes extensions that are referred to as **lenses**. A lens is a collection of additional best practices related to a particular domain or type of solution, such as a SaaS application or an IoT solution. These lenses help architects within their domains to critically analyze their solutions, though the guidance within doesn't broadly apply to all kinds of solutions, such as the main body of the framework. Our review in this chapter will use a mix of framework questions between the main body and the IoT Lens. Links to both resources are included in this chapter's *References* section. In the next section, we will review our solution using a subset of the questions posed by the AWS Well-Architected Framework.

Reviewing the solution

Before we perform a solution review, let's restate the problem, revisit the target solution, and reflect on what was built in this book. This will help us refresh our memory and contextualize the solution review using the Well-Architected Framework.

Reflecting upon the solution

Our fictional narrative had us working at Home Base Solutions as the IoT architect responsible for designing a new home appliance monitoring product. This product was a combination of a hub device that connects to consumers' home networks and interacts with paired appliance monitoring kits. These kits are attached to consumers' large appliances, such as furnaces or washing machines, and send telemetry data to the hub device. The hub device processes telemetry data, streams it to the cloud to train ML models, and hosts local inference workloads using new telemetry and the deployed models. The following diagram shows how these entities are related in consumers' homes:

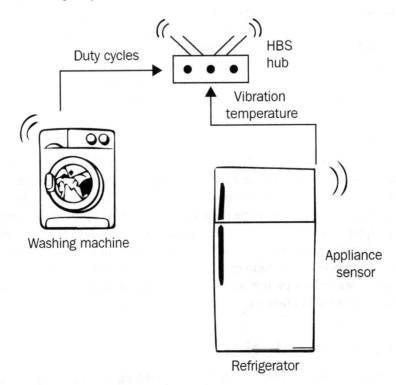

Figure 10.1 – Reviewing the HBS smart hub product's design

Our target solution was to prototype the hub device on a **Raspberry Pi** to collect telemetry data and run the ML workloads, all while using a **SenseHAT** expansion module to collect sensor data and signal results visually to the LED matrix. We used **AWS IoT Greengrass** to deploy a runtime environment to the hub device that could install and run our code as components. These components encapsulated our business logic to collect sensor telemetry, route data through the edge and cloud, fetch resources from the cloud, and run our ML inference workloads.

We used **Amazon SageMaker** to train a new ML model in the cloud using the sensor telemetry that was acquired by the hub device and streamed it to the cloud as training data. This ML model was deployed to the edge to intelligently assess the health of our monitored appliance and signal to the consumer if any anomalous behavior is detected. Finally, we planned how to scale up our solution to a fleet of hub devices, their monitoring kits, and ML models, and how to operate this fleet in a production environment. The following diagram reviews our solution architecture:

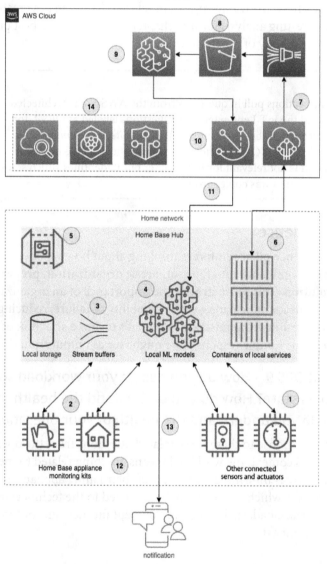

Figure 10.2 – The original solution architecture diagram from *Chapter 1, Introduction to the Data-Driven Edge with Machine Learning*

With this brief review of our business objective and solution architecture to set the context, let's apply the AWS Well-Architected Framework to analyze our solution.

Applying the framework

The format we will use for the Well-Architected review is to state a question from the framework and then respond with an answer in the role of the HBS IoT architect. As a reminder, highlights from the framework were selected from the base material and the IoT Lens to drive interesting analysis for this chapter. There are more best practices to consider in the complete body of the framework.

> **Note**
>
> The following sections pull in questions from the AWS Well-Architected Framework and the IoT Lens extension. Questions labeled as *OPS 4*, for example, indicate that they are from the *Well-Architected Framework*. A question labeled as *IOTOPS 4* indicates it is from the *IoT Lens extension*. This distinction is not relevant for this chapter but it identifies which source material the question was copied from.

Operational excellence

The **operational excellence** pillar reinforces thinking about how we operate the live solution. It organizes its guidance into four sub-areas: **organization**, **preparation**, **operation**, and **evolution**. This pillar stresses the importance of an organization's work culture and mechanisms for anticipating the inevitability of failure, reducing the influence of human error, and learning from mistakes. Now, let's review a selection of the questions from this pillar and some sample responses we might see as output from the architect.

OPS 4, OPS 8, and OPS 9 – How do you design your workload so that you can understand its state? How do you understand the health of your workload? How do you understand the health of your operations?

We will summarize a response to these three related questions from the operational excellence pillar. In this context, the workload means anything related to meeting our business objectives, such as informing customers of their failing appliances. This is different than operations, which refers to anything related to the technical implementation we use to operate the workload, such as the deployment mechanisms or tools we use to notify our team of the impact.

We have designed each level of our workload to report some kind of health state. Our workload can be defined at three levels, each with mechanisms for reporting its state so that we can automate monitoring and alerting. These three levels are the fleet of devices, the components running on a hub device, and the cloud pipeline of training ML models. At the fleet level, hub devices report the health of their deployments and connectivity status to the cloud with services such as AWS IoT Greengrass and Amazon CloudWatch. We can use services such as **AWS IoT Device Management** to monitor for devices in unhealthy states and take action against them. The components that are running on devices are monitored by the IoT Greengrass core software, and logs for each component can be shipped to the cloud for automated analysis. The ML training pipeline reports metrics on training accuracy so that we can measure the overall state of meeting our business objectives.

We will implement threshold alarms on critical failures, such as devices failing deployments and appliance monitoring kits losing connection to their hub devices. These enable us to proactively mitigate failures before they impact our customers, or reach out to customers to inform them of actions they can take to restore local operations.

OPS 5 and OPS 6 – How do you reduce defects, ease remediation, and improve the flow into production? How do you mitigate deployment risks?

To reduce defects and mitigate deployment risks, we must include a physical copy of each target hardware profile running our solution in our testing and deployment pipeline. These devices will be the first to receive new deployments through Greengrass by specifying them as a separate group in AWS IoT Core. We can configure our CI/CD pipeline to create new deployments for that group and wait for these deployments to be reported as successful before advancing the deployment to the first wave of production devices.

We get some out-of-the-box remediation value from Greengrass anyway because, by default, it will roll back failed deployments to the previous state. This helps minimize the downtime of production devices and instantly signals to our team that something is wrong with the deployment. Greengrass can also stop the fleet of grouped devices from being deployed if a certain portion of them fail their deployment activity.

IOTOPS 3 – How are you ensuring that newly provisioned devices have the required operational prerequisites?

In our solution of using Greengrass, we know what the documented minimum requirements are for running the Greengrass software. We used the **IoT Device Tester** software to validate that our target hardware platform is compatible with Greengrass's requirements and can connect to the AWS service. We should use the IoT Device Tester software to validate any future hardware platforms that we want to use as HBS hub devices.

We should also calculate the necessary additional resources that are consumed by all of our components. For example, if we know that all of our total static resources will consume 1 GB on disk, we know we need at least that much, plus room for storing logs, temporary resources, and so on. Once we have calculated the minimum requirements for our solution, we can add a custom test to IoT Device Tester that can validate that each new hardware target is ready to run our solution.

Security

The security pillar reinforces thinking about how to maintain or raise your workload's security posture, such as protecting access to data and systems. It organizes best practices into the following sub-areas: Identity and Access Management, detection, infrastructure protection, data protection, and incident response. This pillar stresses clearly defining the resources and actors in your workload, the boundaries and access patterns between them, and the mechanisms for enforcing those boundaries.

SEC 2 and SEC 3 – How do you manage identities for people and machines? How do you manage permissions for people and machines?

The identities and permissions for people are managed by AWS Identity Access and Management. Our customers will log into their management app using federated identities from OAuth providers such as Google or Facebook or create new usernames directly with us using **Amazon Cognito**. We will tie Cognito identities to the devices they own and interact with using policies.

Identities and permissions for devices are managed by a combination of AWS IAM and AWS IoT Core. The device-to-cloud identity uses an X.509 private key and certificate registered with AWS IoT Core to establish MQTT connections. This can be used to exchange a certificate for temporary AWS credentials. These temporary AWS credentials are tied to an IAM role that has policies attached to it to determine what the credentials are allowed to do with various AWS services. By using unique private keys on each device, the identity of a device cannot be spoofed by a malicious actor.

IOTSEC 10 – How do you classify, manage, and protect your data in transit and at rest?

At the edge, we can classify data as either runtime data that is derived from sensors or used to deliver business outcomes or operational data that comes from software and system logs. In our current design, we do not handle runtime and operational data any differently in terms of management or protection. Here, we have the opportunity to better safeguard any potential customer privacy data, such as video feeds from connected cameras.

At the edge, any data that's in transit between the components of the Greengrass solution is not encrypted. We use the permissions model of Greengrass's components and interprocess communication (IPC) to protect access to data that's published over IPC. Data in transit between leaf devices and the Greengrass device using MQTT is encrypted over the network using the private key and certificate with mutual TLS.

At the edge, data at rest is not encrypted and instead relies on the permissions of the Unix filesystem to protect access to data. We must ensure we use proper user and group configurations to protect access to data at rest. Here, we have the opportunity to put a validation mechanism in place to alert us if new system users or groups are created or modified. To perform security threat analysis each time, we must add a new component to the solution to check whether it has the proper security in place for data access.

From the edge to the cloud, we should use mutual TLS to encrypt MQTT traffic in transit and Amazon Signature Version 4 to encrypt any other traffic that's exchanged with AWS APIs with the temporary credentials. Data at rest that's stored in AWS services uses the encryption policies of each service. For example, data stored in Amazon Simple Storage Service (S3) can use server-side encryption with AWS-managed encryption keys.

IOTSEC 11 – How do you prepare to respond to an incident that impacts a single device or a fleet of devices?

Our operations team has alarms set on the operational health metrics of the fleet of devices. For example, if a device fails a deployment, the operations team will receive an incident ticket as a notification from the alarm. If a group of devices fails a deployment, we will page our operations team for immediate triaging.

We will author a series of runbooks for anticipated failure events for our operations team to follow as a first response. The first step will be to define the minimum set of runbooks needed before we are comfortable with the first wave of production devices.

IOTSEC 8 – How are you planning the security life cycle of your IoT devices?

We will work with our **ODM** to document the security life cycle from the supply chain of parts, through assembly and delivery to our warehouse for inclusion in the retail packaging. It is important to us that parts such as the central processor, volatile and non-volatile memory, and the **Trusted Platform Module (TPM)**, which houses the private key, are authentic and haven't been tampered with before they are assembled into our product.

All TPMs provided by the ODM for our devices will be associated with a CA that we will register in AWS IoT Core. We will pre-provision each device in the cloud so that the devices can simply connect using their protected credentials and not require any JIT registration process.

Should we identify any device as having a compromised identity, we will assess whether a certificate rotation activity is a sufficient mitigation. If not, we will revoke its certificate in AWS IoT Core to prevent it from exchanging further data and proactively reach out to the customer to start an exchange process.

Reliability

The reliability pillar reinforces that a workload should continue to operate as it was designed and when it is expected to. It organizes best practices into the following sub-areas: foundations, workload architecture, change management, and failure management. This pillar stresses concepts such as failover and healing mechanisms in response to failures, testing recovery scenarios, and monitoring for availability during steady-state operations and after deploying a change.

REL 3 – How do you design your workload service architecture?

We have designed our workload service architecture using a service-oriented architecture and implemented the principles of isolated, decoupled services. We use this architecture to make it easier to design, author, test, and ship code, as well as to minimize the impact that an isolated service experiencing faults will have on the solution. We codify this architecture design using the mechanisms defined by the core Greengrass software and its components.

REL 8 – How do you implement change?

For our edge solution, we use versioned components to incrementally update the software running on our devices through Greengrass deployments. We deploy changes on test devices before rolling those changes out to production devices. Deployments that fail on a device will be automatically rolled back. Deployments that fail to 10% of a fleet will roll back the entire deployment to that fleet.

For our cloud solutions, we use CloudFormation templates and stacks to provision cloud resources and make changes to them. We do not make any changes to the production infrastructure not authored through IaC mechanisms. These changes must be reviewed by a peer on the team before they can be deployed. We can use CloudWatch Metrics and Logs for our provisioned resources to monitor for any unhealthy statuses and roll back CloudFormation changes in the event of operational impact.

IOTREL 3 – How do you handle device reliability when communicating with the cloud?

Our edge ML solutions are designed to operate independently from the cloud. Some features are impacted during periods of network instability, such as publishing failure events to the cloud for customer push notifications to their mobile app. Events, telemetry data, and logs that are destined for the cloud are buffered locally and will eventually get to the cloud once network instability has been resolved. Data that is published to the cloud but does not get an acknowledgment of this will be retried, such as with an MQTT quality of service set to an *at least once* level of service.

When the cloud is trying to communicate with devices, such as when a new deployment is ready to be fetched, we use durable services such as Greengrass, which keep track of devices that are offline and haven't completed a pending deployment activity yet.

REL 11 and IOTREL 6 – How do you design your workload to withstand component failures? How do you verify different levels of hardware failure modes for your physical assets?

(In this case, *component failure* does not mean Greengrass components specifically.) Here, we use a service-oriented architecture to withstand component failures so that any of our custom services should be able to fail without bringing down the entire solution. For example, if the component that reads measurements from the temperature sensor fails, the hub device and edge solution will still be operational, albeit with less accuracy when it comes to detecting appliance anomalies.

There are some components provided by Greengrass that, if failing, could impact multiple outcomes in our solution, such as the IPC messaging bus. If components such as these fail, our custom components will not be able to publish new messages, and receiving components would stop getting new messages to work with. We should update our custom component code, which publishes messages, so that it can buffer messages where we cannot afford to drop messages while IPC is unavailable. We should also study the behavior of Greengrass and its ability to self-recover when a provided function such as IPC is impacted.

If any of our cyber-physical hardware interfaces fail, such as a sensor no longer being able to be read, we would stop seeing values being published over IPC and get error messages in the corresponding software component that uses the sensor. We may be able to triage events like these remotely using uploaded logs. If any of our compute, memory, disk, or network hardware components fail, the entire solution will likely be disabled and require on-premises triaging or the device being exchanged through our customer support program.

Performance efficiency

The performance efficiency pillar reinforces that we strike a balance between consumed resources and the available budget and that we continue to seek out efficiency gains as technology evolves. It organizes best practices into the following sub-areas: selection, review, monitoring, and tradeoffs. This pillar stresses delegating complex tasks for solved problems, planning for data to be at the right place at the right time, and reducing how much infrastructure your team must manage.

PERF 2 – How do you select your compute solution?

Concerning our ML model training needs, we will initially select compute instances on AWS based on our default settings and evaluate whether there are more cost-effective instance profiles to use in our training life cycle through trial and error. Since ML is a differentiator for our consumer product, we want to enable the ML model on our customers' devices within an appropriate **service-level agreement** (**SLA**), such as within one business day after accumulating enough training data to produce an accurate model. As we ramp up our production fleet, we may find value in batching training jobs to maximize the utilization of provisioned compute instances.

Concerning our target device hardware at the edge, we will measure the performance of our full production workload on the prototype device, such as a Raspberry Pi, and iterate toward a production hardware profile based on the overall utilization of the compute device. We want to leave some buffer room in the total utilization in case we deploy new workloads to devices as a future upgrade.

PERF 6 – How do you evolve your workload to take advantage of new releases?

We will monitor new releases from AWS for opportunities to bring in new managed Greengrass components that handle even more undifferentiated heavy lifting for our edge workload. We will also monitor new releases in the Amazon SageMaker and **Amazon Elastic Cloud Compute** portfolios for opportunities to optimize our ML training pipeline.

PERF 7 – How do you monitor your resources to ensure they are performing?

We will use the managed component for enabling **AWS IoT Device Defender** to collect system-level metrics from each device, such as compute, memory, and disk utilization. We will monitor for anomalies and threshold breaches and act in response to any detected impacts.

IOTPERF 10 – How frequently is data transmitted from devices to your IoT application?

For high-priority business outcomes and operational alerts, such as informing others of a detected anomaly or a drop in sensor values, data will be transmitted from devices to the cloud as soon as such data is available. For other classes of data, such as reporting component logs or sensor telemetry to use in a new ML training job, data can be transmitted in batches daily.

Cost optimization

The cost optimization pillar reinforces how to operate a solution that meets business needs at the lowest cost. It organizes best practices into the following sub-areas: financial management, usage awareness, cost-effective resources, managing demand and supply, and optimizing over time. This pillar stresses measuring the overall efficiency of your cloud expenditure, measuring return on investment to prioritize where next to optimize, and seeking implementation details that can lower costs without compromising on requirements.

COST 3 – How do you monitor usage and cost?

We will use a combination of Amazon CloudWatch for metrics and logs, as well as the AWS Billing console to monitor the usage and cost of consumed AWS services. The most significant source of cost is anticipated to be cloud compute instances for our ML training workloads. We will monitor the costs associated with each device for outliers where individual devices are consuming more in cloud costs than the fleet's average.

IOTCOST 1 and IOTCOST 3 – How do you select an approach for batch, enriched, and aggregate data that's delivered from your IoT platform to other services? How do you optimize the payload's size between devices and your IoT platform?

To capture sensor telemetry from our appliance monitoring kits, we will batch the telemetry data for daily transmission to the cloud, which will go directly to Amazon S3. This will dramatically lower the cost of the transmission compared to sending each payload as it is published by the sensor components. We do not have plans to further optimize the payload sizes for any operational messages that are exchanged between Greengrass devices and the cloud because we do not anticipate these messages to make up a significant expense.

That concludes our sample responses to the AWS Well-Architected review. Are there any responses you disagree with or would otherwise modify? The review process is a guideline and is not designed to contain right or wrong answers. It is up to you and your team of collaborators to define how complete the answers should be and whether or not you have action items as a result of the review. Questions that the team has no answer to or cannot articulate a detailed answer to are good opportunities to learn more about your architecture and anticipate problems before they surface in your solution. In the next section, we will provide some final coverage of the AWS features you may find useful but that did not fit in the scope of this book.

Diving deeper into AWS services

This book focused on a specific use case as a fictitious narrative to selectively highlight features available from AWS that can be used to deliver intelligent workloads to the edge. There is so much more you can achieve with AWS IoT Greengrass, the other services in the AWS IoT suite, the ML suite of services, and the rest of AWS than what we could cover in a single book.

In this section, we will point out a few more features and services that may be of interest to you as an architect in this space, as well as offer some ideas on how to extend the solution you've built so far to further your proficiency.

AWS IoT Greengrass

The Greengrass features we used in the solution represent a subset of the flexibility that Greengrass solutions can offer. You learned how to build with components, deploy software to the edge, fetch ML resources, and make use of built-in features for routing messages throughout the edge and the cloud. The components we used in the hub device prototype primarily downloaded Python code and launched long-lived applications that interacted with the **IPC** messaging bus. Components can be designed to run one-off programs per deployment, per device boot, or on a schedule. They can be designed to run services that act as dependencies for your other component software, or wait to start once other dependencies in your component graph have run successfully.

Your components can interact with the deployment life cycle by subscribing to notifications about deployment events. For example, your software can request a deferment until a safe milestone is met (such as draining a queue or writing in-memory records to disk), or signal to the Greengrass nucleus that it is now ready for an update.

Components can signal to other components that they should pause or resume their functionality. For example, if a component responsible for a limited resource such as disk space or memory identifies a high utilization event, it could request that the components consuming those resources pause until the utilization comes back into the desired range.

Components can interact with each other's configuration by requesting the current configuration state, subscribing to further changes, or setting a new value for a component's configuration. Returning to the previous example, if a resource watchdog component didn't want to fully pause a consuming component, it could specify a new configuration value for the consuming component to write sampled values less frequently or enter a low-power state.

All three of the previously mentioned features work using Greengrass IPC and are simple applications of local messaging between your components and the Greengrass nucleus that govern the component life cycle. There is lots of utility for these features in your solution design and they demonstrate how you can build systems for component interaction on top of IPC.

Here are a few more features of Greengrass that you should be aware of as you continue your journey as an edge ML solution architect. The documentation for Greengrass's features can be found online at `https://docs.aws.amazon.com/greengrass`:

- **Nucleus configuration**: When installing the Greengrass core software on your device (or later, through deployments of updated configuration), you have several options to explore for optimizing consumed resources, network configuration, and how the device interacts with the cloud service. All of these have intelligent defaults to get you started, but your production implementations may need to include refinements of these per the results of your well-architected review!

- **Run Greengrass inside a Docker container**: In this book's solution, we installed Greengrass as a service running on the Raspberry Pi. Greengrass can also be installed on a device running in its own Docker container. You may find this valuable for simplifying a custom installation across devices using IaC. This can also be used to ship your entire Greengrass solution as an isolated service as part of a grander solution architecture running on the device.

- **Run Docker containers as components**: Your Docker containers can be imported into an edge ML solution without modification if you wrap them as a new component. Greengrass offers a managed component for interacting with Docker Engine running on the device. It can pull images down from Docker Hub and Amazon Elastic Container Registry. This can expedite your path to adopting Greengrass in existing workloads where your business logic is already defined in Docker containers.

Now, let's review a few more features in the wider suite of AWS IoT services that can power up your next project.

AWS IoT services

The suite of AWS IoT services covers use cases for device connectivity and control, managing a fleet at scale, detecting and mitigating security vulnerabilities, performing complex event detection, analyzing industrial IoT operations, and more. Greengrass is a model implementation of designing edge solutions on top of existing AWS IoT services and also natively integrates with them in powerful ways. Here are a few more features in the AWS IoT suite to take a look at when designing your next edge ML solution. The documentation for these features can be found at `https://docs.aws.amazon.com/iot`:

- **Secure tunneling**: In *Chapter 4, Extending the Cloud to the Edge*, you uploaded system and component logs to Amazon CloudWatch to remotely triage your device. What happens if you need more information than what your logs are capturing or you need to run a command on the device but don't want to write a component just for that? With secure tunneling, you can signal your devices to establish SSH tunnels to your operators, eliminating the necessity for inbound network connections to your device. Greengrass has a managed component to enable this feature on your device.

- **Fleet indexing**: When using the shadow service of AWS IoT Core to synchronize the state of your Greengrass devices, leaf devices connected to hubs, and even your components, all of your shadows can be indexed for queries and dynamic groups using the fleet indexing service of AWS IoT Device Management. Fleet indexing makes it easy to search for shadow-backed entities in your solution for analysis and action. For example, you could create a dynamic group of devices reporting under 20% battery level and inform your remote technicians to prioritize battery replacements on those devices.

- **Device Defender**: AWS IoT Device Defender is a service for automating the process of detecting and mitigating security vulnerabilities that can use ML to build a profile of your fleets' normal behavior. The service can inform your security team of devices operating in unusual ways, such as a spike in network traffic or disconnection events that could represent a malicious actor interfering with your device.

Now, let's review a few more services in the ML suite of AWS that add more intelligence to your workloads.

Machine learning services

The ML services of AWS span from tools that help developers train and deploy models to high-level artificial intelligence services that solve specific use cases. While the following services run exclusively in the AWS cloud today, they can help augment your AI edge solutions that can work with remote services:

- **Amazon Lex and Polly**: You can build voice interfaces into your solutions using the same technology that powers Amazon's Alexa voice assistant. Lex is a service for building interactive experiences from voice and text inputs. Polly is a service for translating text to speech. You can use both to process audio requests from your devices and return a lifelike synthesized response.

- **Amazon Rekognition**: You can augment an intelligent workload at the edge with deeper insights from the cloud. For example, your edge ML workload may use a simpler motion or object detection model to capture significant events as video clips, then send only these clips to the cloud for deeper analysis with a service such as Rekognition. This pattern of escalation can help you cut down on the resources that are needed at the edge and reduce the costs of operating ML workloads exclusively in the cloud.

Next, we will provide a few ideas on the next steps you could take to extend this book's solution.

Ideas for further proficiency

With your working solution of a hub device running a local ML workload, you have practiced using all the necessary tools to deploy intelligent workloads to the edge. You may already have a project in mind to apply the lessons you've learned from this book and reinforce what you've learned through practical application. If you are looking for inspiration on the next steps to take, we have compiled a few ideas for extending what you have already built as a means to develop further proficiency with this book's topics:

- **Add a new cyber-physical interface**: The Raspberry Pi's GPIO interface, USB ports, and audio output jack offer a wide design space for extending the cyber-physical interfaces of your hub device. You could add a new sensor, author a custom component for reading values from it, stream values to the cloud, and train your ML model to detect anomalies in that feed. You could plug in a speaker and generate audio alerts in response to the object detection model we deployed in *Chapter 4, Extending the Cloud to the Edge*. You could go one step further and synthesize custom speech audio from Amazon Polly and play that over the speaker!

- **Build a prototype of the appliance monitoring kit**: In this book, we used an onboard sensor from SenseHAT to acquire measurements as an approximation of the appliance monitoring kit. With two devices, you could provision one as the hub device and another as the monitoring kit, connecting the kit to the hub over MQTT, as demonstrated in *Chapter 4, Extending the Cloud to the Edge*. Remove the component from the hub device that emits sensor readings and write a new component that subscribes to an MQTT topic and forwards new messages from the monitoring kit to the same IPC topic, as defined in *Chapter 3, Building the Edge*. Your monitoring kit device doesn't need to deploy Greengrass; it can run an application similar to the one we saw in *Chapter 4, Extending the Cloud to the Edge*, for connecting to the Greengrass MQTT broker using the AWS IoT Device SDK.

- **Perform your own AWS Well-Architected review**: The sample review that we provided in this chapter highlighted a few of the questions that are posed to architects and kept answers at a somewhat high-level analysis. As your next step, you could complete the rest of the review with questions that have not been included in this chapter and also take the opportunity to document your answers to the same questions, perhaps for your workload.

Do you have other suggestions for project extensions or want to show off what you've built? Feel free to engage with our community of readers and architects by opening a pull request or issue in this book's GitHub repository!

Summary

That's all we have for you! We, the authors, believe that these are the best techniques, practices, and tools you can use to continue your journey as an architect of edge ML solutions. While some of the tools are specific to AWS, everything else should generally serve you in building these kinds of solutions. Solutions built with AWS IoT Greengrass can just as easily include components that communicate with your web services or the services of cloud vendors such as Microsoft or Google. The guiding principle of this book was to prioritize teaching you how to build and how to think about building edge ML solutions over using specific tools.

As you take your next steps, whether they are extending this book's prototype hub device, starting a new solution, or modernizing an existing solution, we hope you find value in reflecting upon the lessons you've learned and critically thinking about the tradeoffs that help you reach your goals. We welcome your feedback on this book's content and the technical examples through the communication methods included in this book's GitHub repository. We are committed to the continued excellence of the examples and tutorials provided, recognizing that in the world of information technology, tools evolve, dependencies update, and code breaks.

We are sincerely grateful that you decided to invest your precious time in exploring the world of bringing intelligent workloads to the edge with our book. We wish you the best of luck in your future endeavors!

References

Take a look at the following resources for additional information on the concepts that were discussed in this chapter:

- Infrastructure as Code whitepaper: `https://d1.awsstatic.com/whitepapers/DevOps/infrastructure-as-code.pdf`

- AWS Architecture Center (IoT): `https://aws.amazon.com/architecture/iot/`

- AWS IoT Greengrass documentation: `https://docs.aws.amazon.com/greengrass`

- AWS Well-Architected home page: `https://aws.amazon.com/architecture/well-architected/`

- AWS AI services: `https://aws.amazon.com/machine-learning/ai-services/`

- IoT Atlas: `https://iotatlas.net`

Appendix 1 – Answer Key

This appendix provides the answer key for each set of questions at the end of the book's chapters. Some questions were open-ended in that they did not have an explicit right or wrong answer. For these questions, we have provided sample responses for you to evaluate against your own.

Chapter 1

1. A cyber-physical solution is any interoperation between the analog and digital worlds. An edge solution may include cyber-physical interfaces and also interoperates with other entities in a network topology at a point furthest away from the topology's center of gravity.

2. Automobiles that are now edge solutions have built-in wireless communications for exchanging messages with a remote service for purposes of logging runtime operations and receiving software updates. For example, Tesla Inc. electric vehicles were designed from day one to be intelligent edge solutions in addition to mechanical vehicles.

3. No, the telephone was originally a mechanical device. Despite translating analog audio to electrical impulses, this data was not initially processed with computers. Our *Chapter 1, Introduction to the Data-Driven Edge with Machine Learning*, definition of cyber-physical states that hardware and software are combined to deliver outcomes.

4. Edge solutions have at least one sensor or actuator for interacting with the analog world, some amount of compute power for processing data or instructions, and interact with at least one other kind of entity at a point in time.

5. The three primary types of tools needed to deliver intelligence workloads at the edge are a runtime for orchestrating edge software, a ML model and library, and a method for communicating with other entities.

6. The four key benefits in running machine learning models at the edge are improved latency to process data, improved availability through local autonomy, cost savings from reduced data transmitted to a remote service, and more support for complying with data governance requirements.

7. The primary persona in smart home solutions is the end consumer (homeowner, resident, or otherwise beneficiary of the solution's capabilities).

8. Another potential use case for ML-powered edge solutions in the home is assisted living. People who require additional vital sign monitoring can benefit from the low latency and data governance of local ML.

9. The primary personae of industrial solutions are the operators and maintainers of industrial equipment.

10. Another use case for industrial solutions is optimizing the use of equipment in order to minimize wear and tear, or maximize the output of the equipment. Industrial equipment often needs to run autonomously, meaning an edge ML solution can operate independently of any remote service that could be unavailable when a decision needs to be made.

11. No, typically the IoT architect is responsible for coordinating how ML models are deployed and operated at the edge but is not responsible for their overall accuracy. The IoT architect works with data scientists and product owners to help coordinate overall solution effectiveness, which is bound with the success of properly trained and tested ML models.

Chapter 2

1. The use of isolated services is the best practice for organizing code in edge ML solutions. Monolithic applications may sometimes be the right choice, depending on project requirements.

2. A benefit of decoupling services in edge architecture is limiting the behavioral scope of what those services do. A simple single-purpose service is easier to write, maintain, and reuse.

3. A benefit of isolating your code and dependencies from other services is the assurance that you won't have any dependency conflicts between your services.

4. A key trade-off when evaluating wired and wireless networking is power consumption. Wireless communication needs more power to transmit and receive messages. If the overall solution is wireless, this usually means reliance upon a local battery source for power.

5. A smart home device that uses both a sensor and an actuator is a motorized garage door. These devices use a break beam sensor to detect whether anything is in the path of the door as it is closed by the motor actuator.

6. The two kinds of resources defining an IoT Greengrass component are the recipe and the artifacts.

7. False – a component does not require an artifact. All of the logic of the component could be contained in the recipe life cycle scripts.

8. IoT Greengrass defaults to using the root system user for storing and executing components so that other system users cannot interact with your component files. This helps ensure that your code is running untampered on the device.

9. True – components can be deployed to IoT Greengrass devices locally or through remote deployments. Local deployment is more typical in the prototyping phase of a project.

10. Here are three different methods to update the `Hello, world` component: start a new deployment and merge in the new component configuration overriding the message key, update the default configuration in the recipe and redeploy, and update the script artifact to use a new string and redeploy.

Chapter 3

1. The three network topologies common in edge solutions are star, bus, and tree. Hybrid topologies mix and match these topologies in localized areas of the graph. Our hub device uses a hybrid topology in order to communicate with the cloud but otherwise uses a star topology at the edge to connect local devices.

2. False – IoT Greengrass does not operate at layer one of the OSI model, though some custom components may interact with it.

3. A benefit of using a publish/subscribe model for exchanging messages is decoupling cross-awareness of other devices or components in the solution in your solution code. Abstracting away interactivity to topic addresses shifts interactivity away from a device-specific addressing scheme.

4. True – IoT Greengrass can act as both a messaging client (to the cloud) and a messaging broker (to local devices and components).

5. The `{ "temperature": 70 }` message is an example of structured data and is serializable.

6. Image data from a camera is an example of unstructured data. It is not serializable and would be transmitted as binary data.

7. The worst-case scenario of a compromised home network router is a man-in-the-middle attack where your personal data could be exfiltrated or tampered with. Because the traffic appears to be processed as normal, this could mean the attack remains undetected.

8. A mitigation strategy for verifying authenticity between two network devices is a public key infrastructure.

9. It is important to protect root access on a gateway device so that protected resources and code are not tampered with.

10. A downside to using containers for every component in an edge solution is an increase in the overall disk and memory space. It may not meet your device requirements to use containers in every scenario.

11. IoT Greengrass provides a built-in service called **Interprocess Communication (IPC)** to let components exchange messages.

12. One way to make the sensor and actuator solution more secure is to restrict the allowed IPC topics to just the specific topics needed for the solution instead of using wildcards.

13. To redesign this solution with a third component that will interpret sensor results, this new component would subscribe to the sensor's publishing topic and publish commands to a new topic. The actuator component would then need to be updated to subscribe to this new topic instead.

Chapter 4

1. Static resources are those that don't change across deployments to multiple devices and don't change after deployment. For example, code artifacts are a static resource that should not be altered after deployment or altered per device in a single deployment. Dynamic resources are fetched at deployment, install, or runtime, and may be different each time they are fetched or locally altered after fetching. For example, the customer-specific configuration for a smart home device is a dynamic resource that would be fetched and vary from device to device.

2. IoT Greengrass component artifacts are stored in **Amazon Simple Storage Service (Amazon S3)** for reference in recipe files.

3. You cannot modify an artifact stored in the cloud after it has been registered in a component. This would break the computed digest and flag to IoT Greengrass that the artifact is not safe to use.

4. You can't write over artifact files after deployment since these are considered static resources. There is an expectation that these artifacts remain unchanged since a new deployment would replace any alterations. Deployment of artifacts should be idempotent.

5. True – devices can belong to multiple thing groups and each thing group can define one active deployment.

6. A use case for a device to receive deployments from multiple thing groups is to apply common components across all devices (such as logging or health analysis) and specific components to subsets of devices.

7. True – a single component can both reset configuration and merge in a new configuration.

8. The Moquette component deploys a local MQTT broker for connecting leaf devices.

9. The device shadows component synchronizes the state between the edge and the cloud.

10. The MQTT bridge component relays messages between communication channels such as MQTT, IPC, and the cloud.

11. The IoT sensor publishes messages to an MQTT topic. The MQTT bridge relays this to an IPC topic. The ML inference component subscribes to the IPC topic and analyzes data for alarm states. The ML inference component publishes new alarm messages to an IPC topic. A component that has access to the audio out channel of the device subscribes to alarm messages and plays sounds based on the alarm received.

Chapter 5

1. False – data modeling is applicable for all kinds of databases.

2. The benefit of performing a data modeling exercise is to select appropriate storage solutions such as a database and a schema.

3. The relevance of ETL architectures for edge solutions is that we can define multiple data processing paths based on the data's velocity (Lambda architecture). For example, an edge solution can inspect individual sensor measurements to detect alarming peaks while forwarding measurements in bulk to the cloud for cheaper storage.

4. False – Lambda architecture is a pattern distinct from the Amazon Web Services offering of the same name.

5. One benefit for data processing at the edge is to perform data cleansing and filtering steps on noisy analog data close to the source before incurring costs of transmitting and storing data on the cloud, where the data may not be used at all.

6. The minimum component of Greengrass needed to run is the nucleus.

7. False – managing streams for real-time processing applies to the edge and the cloud.

8. To persist data at the edge for a longer duration of time, configure a data pond solution with the cheapest kind of storage available, such as a local disk.

Chapter 6

1. Two benefits of domain-driven design for edge workloads are as follows:

    - Maintaining a high bar of data quality as data moves throughout a multi-layered system

    - Minimizing developer overhead by making it easier for teams delivering solutions to understand and be fluent in the contexts they are working in

2. False – bounded context and ubiquitous language are distinct concepts within domain-driven design.

3. The benefit of having an operational data store versus a data lake or data warehouse is that specific events and transactions can be indexed and queried in an operational data store. If all transactional data is stored in a data lake as raw input, there must be some additional layer to map this input to the events it represents.

4. A Lambda design pattern brings together streaming and a batch workflow.

5. The data lake strategy can be used to transform raw data in the cloud.

6. False – managed services wrapping a NoSQL database provide APIs to read and write data.

7. The mediator topology is used for scenarios where a chain of steps is guided by a central coordinator. The broker topology is used for scenarios where data producers and consumers can more freely exchange data and react to events.

8. One benefit of using serverless functions for processing IoT data is that only compute resources are consumed when there is IoT data to process. Another benefit is scaling up compute resources to meet the scale of surges of IoT data.

9. Making data available to end consumers can be done with **Business Intelligence (BI)** tools such as Amazon QuickSight, Microsoft Power BI, and Tableau.

10. False – JSON is not the most optimized data format for big data processing. Better examples are Apache ORC and Parquet.

11. An API interface on top of an operational data store could be a serverless HTTP API that accepts parameters for date ranges, device IDs, and sensors, and then runs an optimized query on the data store to return results.

Chapter 7

1. False – there are more than two types of ML systems.

2. The four types of ML systems are supervised, unsupervised, semi-supervised, and reinforcement learning. They range in utility based on how much we know about the input data, and whether we are training for a specific result or looking to see what the algorithm can find in the noise.

3. False – K-means is a clustering algorithm.

4. The three phases of the ML project lifecycle are data collection, data preparation, and modeling.

5. Two frameworks for training ML models are TensorFlow and MXNet.

6. AWS IoT Greengrass is the tool to deploy trained models from the cloud to the edge.

7. False – Greengrass offers managed components ready for image classification.

8. One anti-pattern for ML and IoT workloads is setting an expectation that a single person has the expertise for data preparation, ML training, and deploying solutions to the edge.

Chapter 8

1. A strategy for developing IoT workloads with more agility is to use DevOps methodology.

2. False – DevOps is a methodology first from which new tools are designed to support the benefits of faster software delivery with higher quality.

3. Two challenges for DevOps in IoT workloads are as follows:

   - The edge devices are in remote locations, often outside of our direct control.

   - We cannot throw away edge environments broken by a bad change and start over as we can with a virtual machine in the cloud.

4. Tools such as AWS IoT Greengrass, CloudFormation, and Terraform are used to design DevOps workflows between the edge and the cloud.

5. True – running code at the edge in containers and in AWS Lambda functions offers similar benefits (though each has distinctly unique benefits, too).

6. Three benefits of using MLOps with IoT workloads are productivity (faster iterations on training and deployment), reliability (using CI/CD practices improves the quality of deployments), and auditability (enabling end-to-end visibility of input and output).

7.   The different phases of an MLOps workflow are design, model development, and operations.

8.   True – the ecosystem of MLOps and related toolchains are newer and not as mature as those in the DevOps ecosystem.

9.   AWS provides a SageMaker Edge Manager service for performing MLOps at the edge.

Chapter 9

1.   False – device registration is the process of assigning an identity while device activation is when it comes online for the first time.

2.   A certificate authority can be used with Greengrass to provision certificates for fleets of devices. I can bring my own certificate authority, use one from AWS, or use one from a third-party vendor.

3.   Devices can be provisioned in real time with *just-in-time* processes such as fleet provisioning and just-in-time provisioning using AWS IoT Core.

4.   False – in addition to metrics and logs, there are further monitoring techniques, such as a heartbeat.

5.   A benefit of using a dashboard to view an entire fleet of devices is to identify outliers and take action on them in a single place, speeding up the time taken for recognition and resolution.

6.   A mitigation strategy for remote troubleshooting is to set up reverse tunneling sessions so that operators can interact with devices over the network without compromising on the security of preventing inbound connections.

7.   AWS IoT Greengrass provides managed components such as LogManager, CloudWatch, and the nucleus telemetry emitter.

8.   False – aggregation of metrics can be performed on the edge device. It is not required to be done in the cloud.

Index

Packt.com

Subscribe to our online digital library for full access to over 7,000 books and videos, as well as industry leading tools to help you plan your personal development and advance your career. For more information, please visit our website.

Why subscribe?

- Spend less time learning and more time coding with practical eBooks and Videos from over 4,000 industry professionals

- Improve your learning with Skill Plans built especially for you

- Get a free eBook or video every month

- Fully searchable for easy access to vital information

- Copy and paste, print, and bookmark content

Did you know that Packt offers eBook versions of every book published, with PDF and ePub files available? You can upgrade to the eBook version at packt.com and as a print book customer, you are entitled to a discount on the eBook copy. Get in touch with us at customercare@packtpub.com for more details.

At www.packt.com, you can also read a collection of free technical articles, sign up for a range of free newsletters, and receive exclusive discounts and offers on Packt books and eBooks.

Other Books You May Enjoy

If you enjoyed this book, you may be interested in these other books by Packt:

Hands-On Edge Analytics with Azure IoT

Colin Dow

ISBN: 9781838829902

- Discover the key concepts and architectures used with edge analytics
- Understand how to use long-distance communication protocols for edge analytics
- Deploy Microsoft Azure IoT Edge to a Raspberry Pi
- Create Node-RED dashboards with MQTT and Text to Speech (TTS)
- Use MicroPython for developing edge analytics apps
- Explore various machine learning techniques and discover how machine learning is related to edge analytics
- Use camera and vision recognition algorithms on the sensory side to design an edge analytics app
- Monitor and audit edge analytics apps

Developing IoT Projects with ESP32

Vedat Ozan Oner

ISBN: 9781838641160

- Explore advanced use cases like UART communication, sound and camera features, low-energy scenarios, and scheduling with an RTOS

- Add different types of displays in your projects where immediate output to users is required

- Connect to Wi-Fi and Bluetooth for local network communication

- Connect cloud platforms through different IoT messaging protocols

- Integrate ESP32 with third-party services such as voice assistants and IFTTT

- Discover best practices for implementing IoT security features in a production-grade solution

Packt is searching for authors like you

If you're interested in becoming an author for Packt, please visit `authors.packtpub.com` and apply today. We have worked with thousands of developers and tech professionals, just like you, to help them share their insight with the global tech community. You can make a general application, apply for a specific hot topic that we are recruiting an author for, or submit your own idea.

Share Your Thoughts

Now you've finished *Intelligent Workloads at the Edge*, we'd love to hear your thoughts! Scan the QR code below to go straight to the Amazon review page for this book and share your feedback or leave a review on the site that you purchased it from.

https://packt.link/r/1-801-81178-4

Your review is important to us and the tech community and will help us make sure we're delivering excellent quality content.

www.ingramcontent.com/pod-product-compliance
Lightning Source LLC
Chambersburg PA
CBHW062049050326
40690CB00016B/3032